THE BATH & WEST

The Bath & West

A BICENTENARY HISTORY BY KENNETH HUDSON

with a Foreword by Sir Emrys Jones

MOONRAKER PRESS

The Society would like to record its appreciation of the generosity of the following in undertaking to purchase a substantial number of copies of the book, and in this way to provide a guarantee which has made the publication of the work possible:

ATLAS DISPLAY (TENT-HIRE) LIMITED
CRICKET MALHERBIE LIMITED
ESSO PETROLEUM COMPANY LIMITED
H.T.V. WEST LIMITED
THE MIDLAND BANK LIMITED
THE NATIONAL WESTMINSTER BANK LIMITED
THE N.F.U. MUTUAL INSURANCE SOCIETY LIMITED
SHOWERINGS LIMITED
YEO BROTHERS PAULL LIMITED

© 1976 Moonraker Press
26 St Margarets Street, Bradford-on-Avon, Wiltshire
SBN 239.00156.7

Text set in 11/13 pt Photon Times, printed by photolithography, and bound in Great Britain at The Pitman Press, Bath

Foreword

BY SIR EMRYS JONES, PRINCIPAL OF

THE ROYAL AGRICULTURAL COLLEGE, CIRENCESTER.

Scotland can claim to have been the pioneer in the formation of farming clubs. The first known agricultural society appears to have been 'The Honourable Society of Improvers in the Knowledge of Agriculture in Scotland' founded in 1723, with headquarters in Edinburgh. This was followed by the formation of other societies in Scotland, England and Wales, including the Bath Society in 1777.

This was the period when agricultural technology took a mighty step forward. It was the era of the enclosure movement, which replaced the open-field system with convulsive rapidity. The new structure of British agriculture that emerged facilitated the adoption of the new ideas and inventions of the pioneers of the latter half of the eighteenth century. Parliament was predominantly composed of members drawn from the landed gentry, and between 1734 and 1832 fully three-quarters of the members of the House of Commons had a vested interest in land. The farmers themselves were utterly oblivious of the need of any kind of political action and it was taken for granted that, with the Squire in Parliament, their political interests would be safeguarded and all they had to do was to farm their land as best they knew how. The earliest form of association were gatherings of landowners and farmers to discuss methods of husbandry, new systems of farming, new crops and livestock improvement. At this time Coke of Holkham was prominent in this kind of activity and his 'Coke's Clippings' became large gatherings, attracting farmers and landowners from far and wide. Agricultural historians tend to concentrate on the developments in Norfolk and the East Midlands in the latter half of the eighteenth century, and for this reason there are few students who are unaware of the achievements and influence of Jethro Tull, Lord Townshend and Robert Bakewell.

Yet in the last two decades of the eighteenth century the Bath Society as it then was had initiated a comprehensive range of experiments and recorded the results in a most meticulous manner. This insistence on accuracy and attention to detail is revealed in the words of a member of the Society who wrote as early as 1780, 'Useful hints of the speculative kind, which may in their consequences lead to practical improvements, have not been neglected;—such will always be esteemed valuable communications, although inferior to those that have already been submitted to test and experiment.' This critical approach to the recording of novel experiences in the field undoubtedly contributed much to the exact scientific method which was to develop later in field experimentation.

This book describes the events that led to the foundation of what was to become the Bath & West Society and records the history and growth of that Society up to modern times. It spans a period of two centuries, when British farming experienced the most far-reaching changes. The author reveals in a most vivid and lucid manner the ways in which the Society stimulated the improvement of agriculture in the South West. By the 'diffusion of knowledge, both scientific and practical,' and by the encouragement of research and education in improved husbandry practices, its impact on farming in the South and West of England was tremendous. Indeed the Bath & West Society had followed the approach of 'Practice with Science' for over half a century before the foundation of the Royal Society of England!

Contents

Acknowledgements

The preparation of this History has been watched over in the most friendly and easy-going way since 1969 by the Society's Library and Publications Committee, which has had to face problems of steadily increasing costs in book-publishing and of the sheer bulk of information and opinion generated by two hundred years of activity. The wide range of temperaments, ages and interests represented on the Committee has been a great help, since it has acted as a constant reminder of the fact that the people who might be expected to buy and read this book would not be all of a piece. The discussions, and sometimes arguments, with the members of the Committee, in and out of meetings, have been of enormous value to me. Among the many people who have given me help and encouragement, I should like to express my gratitude especially to Sir Gerald Beadle, who did a great deal to get the project off the ground in the first place; John Davis, whose formidable knowledge of the Society's history and personalities has been exceeded only by his remarkable tact and discretion; Philip Bryant, the Society's Honorary Librarian, who has pointed out many items of information that I should otherwise have missed; Andrew Jewell, who most kindly placed the resources of the unequalled library at Reading at my disposal and who has always been ready with expert advice.

The Chief Executive, Lord Darling, always resists the slightest attempt to compare him with the great Secretaries of the past, but he will receive his proper due in the tercentenary History, to appear in 2077, when his achievements in developing the Shepton Mallet ground will be seen full-sized and in their historical context. Meanwhile, one can do no more than thank him for his never-failing kindness and ever-welcome common sense, which have made unlikely things happen and prevented a number of errors of tactics.

My wife has almost certainly combed the Society's archives more thoroughly than they have ever been combed before. Her keen eye for significant details and her understanding of English rural society in the eighteenth and nineteenth centuries have been invaluable, and a great debt of gratitude is due to her, both from the Society and from myself. That she has found so much to interest her in these old records is perhaps some compensation for the hours she spent working her way through them.

Ann Nicholls has prepared the typescript in her usual exemplary fashion, and only those who have had to deal with my handwriting will be able to understand the size of this achievement. To Terry Darvill, who has handled documents, pictures and other precious relics with the greatest photographic skill, I am exceedingly grateful.

Preface

When, in 1967, I was asked to prepare the bicentenary history of the Society, I agreed, with the proviso that it would be necessary first to write a book about British agricultural societies in general—a subject, curiously enough, which nobody had so far attempted. The first book was published in 1972 by Hugh Evelyn, under the title of Patriotism with Profit. *This survey, in which, not sur-prisingly, the Bath & West figured prominently, ended with the following com-ment on the work of the agricultural societies.*

'Their aim was, more or less, to make rural people think and to enlighten their minds. Education is always an extremely uncertain process and those who set out to achieve it often met with curious surprises. The apparently intelligent turn out to be buried in their set ways and the apparently stupid to be capable of showing an interest in the most unlikely subjects. The important thing, however, is to persist, despite all discouragement, in laying new ideas in front of people, and to reckon that, if one's pupils fail to show any interest, there must be something unsatisfactory about one's own methods of putting the ideas across. Conservatism is the guiding feature of most people's lives—country people have no monopoly of it—and it takes an unusual degree of effort and imagination or possibly of desperation, to decide to adopt a new technique, a new philosophy or a new attitude towards one's work or one's pattern of living.

'The successful innovators within the agricultural societies were the men, and occasionally the women, who combined an unquenchable enthusiasm for progress with a sensitivity to the hopes and fears of the farming community. This is another way of saying that genuine progress is achieved by people of excep-tional humanity, exceptional energy and exceptional imagination. Perhaps a dozen of the pioneers and enthusiasts we have mentioned in the preceding pages had these qualities. Without them, it is doubtful if either the Bath & West or any other provincial society would have achieved a great deal.'

In 1973 this theme was amplified in my lecture, The Four Great Men of the Bath and West—*the first in the Society's new series of biennial lectures—when I decided to confine my choice of Great Men to people who were long and safely dead. The problem of dealing tactfully and fairly with living notabilities is almost insoluble, and in the present book I have avoided, rather than attempted to solve it, by a final chapter which does little more than summarise the facts and allows the protagonists to do a little talking on their own behalf.*

I make no apology for having quoted so liberally throughout the book from the

Society's own publications, and especially from the volumes of the Letters and Papers *and the* Journal. *So much interesting material is to be found there, and so much good writing, that it would have been foolish and unfair to the reader to have paraphrased it. The flavour of these articles is part of the story.*

The Society's Library and Archives

In 1864 C. P. Russell, Librarian of the Bath Royal Literary and Scientific Institution, compiled a catalogue of the Society's books, which at that time were being looked after by the Institution. 'Now, exactly a century later,' reported the Chairman of the Library Sub-Committee,[1] 'there has been a complete overhaul and reorganisation of the Library. This has been effected through a Sub-Committee formed in 1960, which included Mr Arthur Duckworth and Professor T. Wallace, CBE, MC, VMH, FRS, former Director of Long Ashton Research Station, also Mr F. C. Hirst, FLA, Librarian, Ministry of Agriculture, Fisheries and Food. To this Sub-Committee Mr Peter Pagan, Director of the Bath Municipal Libraries, and Mr Philip Bryant, a senior member of his staff, have acted as Honorary Librarians. The Society is greatly indebted to them for their able services.'

Behind this statement lies an interesting story. In 1958 Professor L. P. Pugh came to the Society's offices in search of a copy of a rare book, the *St. Bel Plan for establishing an institution to cultivate and teach veterinary medicine*, which he required in connection with a book he was preparing and which appeared in 1962 under the title, *From Farming to Veterinary Medicine, 1785–1795.*

The *St. Bel Plan* proved difficult to trace and the Secretary, Mr Yardley, decided that the time had come to do something fairly drastic about the Library. He sought the advice of the then Director of Bath Municipal Libraries, Peter Pagan. Following a detailed report by Mr Pagan, the decision was taken to refurbish the collection, which had fallen into a very bad state of repair.

Between 1777 and 1964 the Library had suffered a good deal from several moves and long-standing neglect. Books had been pilfered, lost and damaged and a number sold off. Since 1964 cleaning and rebinding have greatly improved the general condition of the books, but one or two of the more valuable items have been sold to raise much needed capital. In 1972 it was agreed that the purchase of new books should be 'restricted to those relevant to the interests of the Society'.

In preparation for the move of the Society's headquarters from Bath to Shepton Mallet in 1974, the bulk of the Library was transferred to the University Library in Bath, where it is kept under specially secure conditions. The archives—18 volumes of these had already been rebound and repaired—were temporarily taken to Shepton Mallet, but in February 1975 all but the most recent material found a more satisfactory home in the Records Department of Bath Corporation, in the Guildhall. These early records are, however, still the property of the Society.

The first nineteen volumes of the Archives, containing minutes and correspondence from 1777 onwards, have been microfilmed, and one of the microfilm copies can be consulted at Shepton Mallet, where there is also a complete set of the *Journal*.

[1] D. M. Wills. Foreword to the 1964 *Catalogue of the Library*.

List of Illustrations

CHAPTER ONE

BRITISH AGRICULTURE
IN THE SECOND HALF OF THE EIGHTEENTH CENTURY

Until roughly twenty-five years ago, the far-reaching changes which took place in British farming during the eighteenth and early nineteenth centuries were commonly held to have been caused by three things: the enclosure of common land; the introduction of new crops and implements; and the activities of a small body of pioneers. Lord Ernle, writing before the First World War and at a time when the Great Man interpretation of history was still fashionable, was able to declare without hesitation that the progress which became so marked between 1760 and 1830 could be 'broadly speaking, identified with Jethro Tull, Bakewell of Dishley, and Coke of Norfolk'.[1] These, in his view, were the men who gave Britain new crops, new ideas about stockbreeding, new ploughs, drills and hoes and, perhaps most important, a proper respect for farm book-keeping. Their original ideas gradually caught on among other wealthy landowners with a taste for experiment and afterwards percolated slowly downwards to the general mass of farmers. Meanwhile, the enclosure movement, helped along by a long series of Acts of Parliament, was pushing out inefficient, obtuse peasants and clearing the way for men with the capital and the drive needed to bring Britain's underused acres to a more satisfactory level of production, so enabling a rapidly increasing urban population to be fed. All this added up to the Agricultural Revolution.

Recently, this view has been shown to be over-simplified, over-personalised and not in accordance with the evidence. The Agricultural Revolution turns out after all to be as grandiose and misleading a term as its companion, the Industrial Revolution. Change was more gradual and less dramatic than the Lord Ernle generation wished to believe. W. G. Hoskins, for instance, found abundant evidence to prove that the old open-field system had been considerably modified much before the period of Parliamentary enclosure,[2] R. A. C. Parker made clear that the 'Norfolk system' was established on his estates at Holkham long before his time,[3] J. D. Chambers showed that many small owners had been bought out well before enclosure took place,[4] and W. E. Tate considered, in direct opposition to the traditional view, that the opposition to enclosure was not great and that the legal claims of small owners were fairly met.[5]

The most important re-interpretation of the agricultural revolution, however, is found in the work of E. L. Jones.[6] Jones emphasises that the beginnings of agricultural improvement in Britain lie a good deal further back than at one time

1

believed, and that much of the credit in both the 17th and 18th centuries should be given to pioneers who are shadowy, if not anonymous. The changes, in Jones' view, were brought about mainly by a number of little-known farmers and experimenters, rather than by three or four men who have received more than their fair share of publicity and praise.

In agriculture, as in any other branch of human activity, to have a new idea is one thing, but to get it applied and generally accepted is quite another. Coke reckoned that his improvements travelled at the rate of a mile a year. As one writer commented in the 1780s:

'Improvements in Tillage arise, in general, from the slow operation of doubting experience among men who obtain their bread by the sweat of their brows, whose minds are not sufficiently enlarged to adopt, but with reluctance, any deviation from the practice of their forefathers, and who are fearful of risquing the moderate certainty they possess for the prospect of greater gains which are yet unknown.'[7]

Farmers are, in other words, a conservative lot, and it could be argued that the essence of the Agricultural Revolution lay not so much in the discovery of new techniques but in convincing the generality of farmers that these techniques could bring them a higher and more reliable income. The obstacles in the way of progress were very great and they were overcome quite as much by the initiative and perseverance of a few dozen local agricultural societies as by the genius and flair of the handful of Great Men whose names can be found in every history book. The early agricultural societies were established, from the 1730s onwards, in the belief that they provided the best means of making the most advanced farming practice known to the greatest number of people. Many of these societies lasted no longer than ten years, but the widespread eagerness to get them established illustrates the spirit which characterised the best farming and the most lively-minded farmers during the second half of the century.

By the beginning of the nineteenth century, a new breed of farmer was clearly in being. Arthur Young recorded that 'when I passed from the conversation of the farmers I was recommended to call on to that of men whom chance threw in my way, I seemed to have lost a century in time or to have moved a thousand years'.[8] The local agricultural societies are entitled to much of the credit for the change from the old world to the new which so impressed Arthur Young.

Getting the Society launched

'In the Autumn season of the year 1777, several gentlemen met at the City of Bath, and formed a Society for the encouragement of Agriculture, Arts, Manufactures and Commerce, in the Counties of Somerset, Wilts, Glocester, and Dorset, and in the City and County of Bristol.'[9]

The meeting, held at York House on September 8, was attended by twenty-two people. They had come together in response to an advertisement in the *Bath Chronicle* and certain other newspapers, which appealed to 'public-spirited Gentlemen' to give their 'countenance and protection' to such a society, and their names are recorded in the Minutes.

'John Ford Esqr in the Chair,

Revd. Dr. Wilson	Phillip Stephens, Esq.
Revd. Mr. Ford	Paul Newman, Esq.
Dr. Wm. Falconer	Mr. John Newman
Dr. Patrick Henley	Willm. Street, Esq.
Wm. Brereton, Esq.	Mr. Symons, Surgeon
Mr. Saml. Virgin	Mr. Crutwell, Surgeon
Mr. Richard Crutwell	Mr. Arden
Mr. Foster, Apothecary	Mr. Wm. Mathews
Mr. Cam Gyde	Mr. Parsons
Mr. Benj. Axford	Mr. Edm. Rack
	Mr. Bull'

The most distinguished of the founder-members was certainly Dr. Falconer, physician and surgeon, Fellow of the Royal Society, author of many books on medicine, science, religion, politics and classics, and a man of remarkable attainments and versatility. After retiring from practice in London, he settled in Bath and became Physician to the General (now the Royal United) Hospital. He lived in the Circus and died there, still a member of the Society, in 1824. Of the others, one should draw particular attention to Edmund Rack and William Matthews, both Quakers, and the Society's first and second Secretaries respectively, and to Richard Crutwell, the Bath printer and publisher, whose help and influence were important for many years.

At this first meeting, those present were told about the aims and methods of other agricultural societies already in existence[10] and eight resolutions were then passed.

'1. That the Gentlemen now met together with the Noblemen and others whose

3

names are entered in the Subscription Book, do constitute a Society for the above-mention'd purposes, and that the said Society is by this resolution established.

2. That it is necessary immediately to appoint a Secretary to transact the business of the Society: and that it is the unanimous request of this Meeting that Mr. Edmund Rack of Bath do accept the office—Mr. Rack having agreed thereto, is accordingly appointed.

3. That the Thanks of this Meeting be given to Mr. Cam Gyde for the unsolicited offer of his Rooms for the future Meetings of this Society.

4. That the first Meeting of this Society shall be held at Mr. Gyde's Rooms on Fryday the 19th instant at nine o'clock in the Morning, and that the presence of the Nobility and Gentry be, in a circular letter, earnestly requested, to patronize this laudable and honourable Institution.

5. That, previous to the Meeting of the 19th instant, the Members now present shall resolve themselves into one General Committee, to whom the plans and other Materials now produced shall be referr'd, and that the said Committee shall form from the whole, *One General Plan* to be laid before the next Meeting for its approbation: any five of the said Gentlemen to be a Quorum; and the said General Committee shall meet at Mr. Gyde's Rooms on Monday the 15th instant at 11 o'clock in the forenoon.

6. That in order to expedite the business of the above named Meeting, Dr. Wm. Falconer, Mr. R. Crutwell, Mr. C. Crutwell, Mr. Wm. Mathews, together with the Secretary, be a sub Committee to make the necessary previous preparation.

7. That the Secretary, at the request of this Meeting, do immediately form, and cause to be printed and dispersed throughout the above Counties a Sketch of the Institution for general information.

8. That these resolutions together with the names of the present List of Subscribers, be published in the Bath, Bristol, Salisbury, Gloucester and Sherborne Papers.'

The historic Minute was signed by Edmund Rack as Secretary, and the Society was in being.

Rack at that time was forty-two years of age. Most of what is known about his earlier life is to be found in the biographical sketch included in Collinson's *History of the County of Somerset*,[11] to which Rack himself was an important contributor.

'Mr. Edmund Rack,' this tells us, 'was born at Attleborough in the County of Norfolk. He was educated in the religion of his parents, Edmund and Elizabeth Rack, who were both Quakers. We are informed that his father, a labouring weaver, was a man of excellent character; and that his mother was well-known for her preaching, and highly esteemed among her own sect.[12] Thus humble in his parentage, he had little opportunities of instruction at that early season when the mind is best disposed for receiving it. The knowledge of arithmetick was Mr. Rack's highest attainment, when he was removed to Wymondham, as an apprentice to a general shopkeeper, and though possessing talents that disdained the drudgery of his occupation, he was never heard to repine at the necessary labours surrounding it. An employment of this nature must exact that mechanical regularity, which (though common abilities may submit to it without reluctance) is, of all things, most insupportable to genius.

'At the close of his apprenticeship, he went into Essex; and at Bardfield became a shopman to Miss Agnes Smith, whom he married not long after his residence in that place. The servilities of his station were now in some measure done away. Nor were his talents unobserved; for though his employment was in some measure an obstacle to social communication, he had the good fortune to introduce himself to the friendship of a silent few, who contributed to cheer the gloom of his obscurity.'

Rack's business ambitions appear to have been limited to making enough money to allow himself a pleasant life and an early retirement. In 1775, at the age of forty, he felt he had reached a position where he could say farewell to the draper's shop and devote himself to more cultural pursuits, or, as Collinson has it, 'to step forward as a more respectable member of society'. Why, even so, he should have decided to remove himself from East Anglia, where all his roots were, and to migrate to the most fashionable and expensive city in England is not clear. The answer may have centred on his literary ambitions. While he was still in business at Bardfield he had been publishing articles and poems in a number of periodicals and he may well have felt that his talents would flourish better in the social atmosphere of Bath than in the more restricted environment of Bardfield. He may, too, have wanted to make a complete break with commerce and to begin a new life altogether in a place that was associated with enjoying money, rather than making it.

Rack wrote a good deal during his early years in Bath: two volumes of essays and poems,[13] a life of William Penn, the Quaker,[14] and *Mentor's Letters*,[15] a collection of epistles written to guide young people through the rocks and whirlpools of life. All but one of these were published by Richard Crutwell, in Bath. The poems are agreeable, but undistinguished. Rack himself claimed mainly negative virtues for them: they contained no political abuse, no polemic-mongering, nothing, as he put it, 'injurious to the cause of morality, or that can give birth to blush in the conscious cheek of modesty'.

William Penn, 'steady and immovable in the prosecution of wise and noble designs', was an outstandingly noble and heroic figure to Rack; a man whose 'internal peace was founded on the impregnable basis of integrity and substantial virtue', who 'despised the empty noise of popular clamour and the mere whistling of a name'.

Mentor's Letters reflect orthodox Quaker opinion on most of the human pleasures and temptations. Young men are advised to choose their amusements and methods of relaxation with great care and prudence. Racing and bloodsports are particularly to be avoided. 'The cruel diversions of horse-racing, cock-fighting, bull-baiting, etc. etc. are a reproach to any civilised people, and there can scarcely be a greater contradiction than to suppose those to be real Christians, who delight in and practise them.'

Novels and plays were almost equally dangerous, and equally liable to rot and destroy the character. They have, Rack assured his readers, 'been injurious to thousands; they may in general be compared to that species of natural poison, which operates insensibly, while it destroys the vitals of the human constitution; their operation on the mind is not felt but by its consequence, and often too late to be easily remedied. They amuse, soften and enervate the mind; lull it into an in-attention to its important duties; relax the springs of that steady masculine vir-tue, which alone insures peace and happiness; and by degrees, reconciles objects to its view, which the exercise of reason and prudence would discover to be its most dangerous enemies.'

This, one might reasonably think, was a curious man to settle in pleasure-loving, scandal-mongering Bath. Yet there is ample evidence that he was quickly welcomed into local literary circles. These included Lady Miller's poetical revels at Batheaston and Mrs. Catharine Macaulay's coterie at Alfred House. Catharine Macaulay was particularly odd company for Rack to keep. He took a most un-Quakerly part in the celebrations, frivolous by any standard, which took place on her birthday in 1777, composing one of the six odes which were recited in the lady's honour and editing the pamphlet which recorded the proceedings.

There was something else that appears markedly out of character for such a cautious and serious-minded man. Rack, who was an asthmatic and had suffered a severe bout of jaundice, became one of the earliest dupes of the notorious quack, Dr. Graham, who was patronised and encouraged by Mrs. Macaulay. Graham went in for such profitable techniques as placing his patients on 'magnetic thrones' and in electric baths and milk baths. He also prescribed for them his famous and expensive 'aethereal balsamic medicines', which contained the secret of living for 150 years. Rack fell for him.

This, then, was the man who was mainly responsible for getting the Bath Society established. He had a very real interest in agriculture, although he was without any practical experience of it, and he was entirely of his age in the patriotic fervour with which he approached the subject. 'The value of our acres,' he wrote in a survey of agriculture from the time of Adam,[16] 'is the grand source

glutinous Mass. This Compost well incorporated will
as rich as any that can be laid on Land; & 30 bushels ⅌
acre will be a sufficient Quantity.

Were, if Green Weeds collected from our public Roads
(where they are suffered to run to Seed & become a nuisance to
all the adjoining Fields), and Fern from our Heaths, would
not answer the same purpose in inland Counties if mixed
with Lime & Earth in the same Manner? The Mass being
confin'd would ferment much sooner than when laid in
an open Exposure, and much more of the Salts retain'd, which
are the most active principle in Vegetation

Edm⁰ W⁰ Rack,

Sep. 13. 1778.

2. Edmund Rack's recipe for Compost, in his own hand, 1778.

of national riches, and this value will ever bear an exact proportion to their cultivation and produce'. Faced with the need to support increasing populations, Europe as a whole was giving an unprecedented amount of attention to improving its agriculture. 'Even Italy (sunk as it is in luxury and the enervating arts of pleasure) has not been totally inactive', and 'the Hollanders are the only people now in Europe who seem to look upon agriculture with indifference.' But Britain was supreme, and Rack felt able to declare with every confidence that 'England alone exceeds all modern nations in husbandry'. He had written articles for the *Farmer's Magazine* in the same vein.

Edmund Rack, one should remember, came from East Anglia, the county of Coke of Holkham and of Arthur Young, where farming, or at least the best farming, was a good deal more advanced than it was in the West Country. Much of what he saw when he arrived in Somerset must have caused him considerable pain and surprise. Something, he felt, had to be done about it, and, in articles he wrote for the *Bath Chronicle* and for the *Farmer's Magazine*, he set out his ideas for a society, based in Bath, which would devote itself to spreading up-to-date ideas about farming and to persuading and encouraging the farming community to adopt them. To a man of Rack's temperament it was not possible, however, to consider agriculture in isolation. His aim was, in his own words, 'the diffusion of useful information in general' and he envisaged a society of broadly cultured people who would share his wish to advance the welfare of mankind, an admirable and thoroughly eighteenth-century ambition.

In November 1777, a much larger group of people than had assembled at York House two months earlier elected the principal officers of the Society, including the Earl of Ilchester as President. The founding fathers were very local—Lord Ilchester was the only prominent county member—and this fact was emphasised by the original title of the Society. It was the Bath Society and it was thirteen years before it widened its horizons and became the Bath & West of England Society. The original group was gradually supplemented and strengthened by the addition of people of greater influence and expertise, men like Arthur Young, Dr. Priestley the chemist, and Thomas Curtis, the botanist.[17] Lord Ilchester, incidentally, was not a particularly assiduous President. After two years he expressed a desire to resign. The offer was accepted and the Chairman wrote to the noble earl accordingly, at the same time 'requesting that his Lordship would pay his subscription'.

It is interesting to notice how wide and significant an overlap there was between the Society's early membership lists and the subscribers to the volume of poems, essays and letters which Rack brought out in 1781. The 483 subscribers to the poems included many distinguished public figures, among them Thomas Coke of Holkham Hall, Arthur Young and the Duke of Marlborough. Many of the same people, distinguished and undistinguished, were persuaded to support the new Bath Society and the same local printer, Richard Crutwell, printed and published both Rack's literary works and the Society's *Journal*. A further link between the two sides of Edmund Rack's life is the pleasant fact that the Chairman of the Society's first General Meeting was Sir John Miller, of Batheaston, the husband of the lady whose poetic revels have already been noted.

At the first General Meeting, the Society set out its aims:[18]

'To promote the good of community by the encouragement of industry and ingenuity;—to excite a spirit of enquiry which may lead to improvements not yet known;—and to bring speculation and theory to the test of accurate experiment, are the grand ends intended by the present laudable and honourable institution.

'A desire to introduce into the western counties those obvious advantages which the public have reaped in the several parts of this kingdom, where societies of a similar kind have been formed, first excited the idea of the present establishment, in support of which the spirit of *true patriotism* has distinguished itself with unexampled ardor.'

The annual subscription was to be 'not less than One Guinea', with life membership at 12 guineas, and a system of premiums, or prizes, was to be instituted, 'directed to Improvements in Agriculture, Planting, and such Manufactures as are best adapted to these Counties', but 'the Premiums offered in any one

RULES and ORDERS

OF THE

SOCIETY,

INSTITUTED AT *BATH*,

FOR THE

ENCOURAGEMENT

OF

Agriculture, Arts, Manufactures, and Commerce,

IN THE

COUNTIES of SOMERSET, WILTS, GLOCESTER, and DORSET.

BATH:

Printed, by Order of the Society, by R. CRUTTWELL, in St. James's-ſtreet, 1777.

3. Title page of the Society's *Rules and Orders*, 1777.

NAMES OF THE PRESENT MEMBERS

ALPHABETICALLY ARRANGED.

His Royal Highneſs the DUKE of YORK.

Their Graces the DUKES of

BEDFORD
*BUCCLEUGH

MARLBOROUGH
*MONTROSE.

MARQUISSES of
BATH and *LANSDOWN.

The EARLS of

AILESBURY
ARUNDEL
BATHURST
CAMDEN
*CORK
*DARTMOUTH
DIGBY

EGREMONT
FITZWILLIAM
*ILCHESTER
PETERBOROUGH
*SHAFTESBURY
STRAFFORD
*WINCHELSEA.

LORDS

BATEMAN
GLASTONBURY
GWYDIR
*HAWARDEN
*PETRE

*RIVERS
ROLLE
JOHN RUSSELL
SHERBORNE, AND
SOMERVILLE.

BARON DIMSDALE.

Thoſe marked thus * are members for life.

4. Names of the Present Members: the Nobility, 1779.

Year shall not exceed two thirds of the Fund in Hand, at the time the said Premiums are offered by the Society'. Premiums were also to be awarded for 'the best-written and most useful original Essay on any of the subjects to which the views of this Society may be extended'.

The Secretary's duties were spelt out with great precision. He 'shall procure all such Books and Stationary Ware as are needful for the Society's Use, and keep fair Accounts of all Monies received and disbursed by him: The said Accounts to be settled and balanced at each Monthly Meeting in the Society's Cash-book, where a Committee of Accounts shall be appointed to audit them. He shall also perform the necessary Business of his Office with Diligence and Integrity, viz. Attend all Meetings and Committees of the Society;—make all Minutes and Resolutions, and enter them fairly in the Journal or Committee Books;—read all Letters and other Papers sent to the Society, and prepare such Answers thereto as the Society shall direct, and record regularly in the Book of Correspondence such as are worthy of Preservation;—sign all Publications, Notices, and Receipts, and prepare an Abstract or Annual Register of the Transactions of the Society, with particular Accounts of the Improvements made in the various Articles to which its Views are extended, together with such Informations received as appear to be the most useful and important, and the Resolutions of the Society thereon;—attend to the Printing of whatever the Society may direct to be printed, and correct the Press. He shall also collect Subscriptions, and visit Manufactories, or apply for particular Information respecting them when required by the Society so to do; and as much as possible make himself acquainted with the various Arts, &c. &c. to which the Views of this Society shall be directed: He shall also regularly enter the Minutes, Proceedings, and Resolutions of each Meeting, for the Inspection of the next; and that, in Consideration of his Trouble, and the close Attention he must give to this Business, he shall be allowed an annual Salary.'

Rack was in fact paid 50 guineas a year, 'till the Society be better able to increase that sum', together with £30 for 'use of sundry rooms in this house (York House, where he himself lived), Furniture, Fire and Candles'.

A number of other practical matters were decided during the first few months of the Society's existence. One of the most urgent concerned banking arrangements. In January 1778 it was resolved; 'that the money belonging to the Society be deposited in the banks of this City'. There were three banks in Bath at this time—Cam, Whitehead and Phillott; Horlock, Mortimer and Co., and Peach, Kington and Co.—and they were chosen by lot to be Treasurer each in turn for three years. This was a wise precaution, since at this period banks failed quite frequently. By rotating its funds around the banks, the risk was spread, or so the Society believed.

Another important early decision involved the Library. In February 1778 a committee was set up to consider what books should be bought to establish the collection. Its list of the titles and prices of what it proposed to acquire reads:

'*Dr Hunters Evelyns Sylva	2	12	6
*Stillingfleets Tracts		6	0
Du Hamels Tracts	(no price given)		
Farmers Magazine 2 vols.		12	0
*Andersons Essays on Agriculture		12	0
La Kaines Gent Farmer		6	0
*Complete Farmer, or Gen. Dict. on Agriculture	1	5	0
*Universal Gardener & Botanist by Maw			
Abercrombie & Etc.	1	7	0
Everyman His Own Gardiner		5	0
*Rural Improvements		5	0
*Blythes Husbandry		1	6
Humes Essays on Vegetation		2	0
Hartlib on Agriculture		1	6
Hartes Essays		5	0 '

These were all purchased, and to them were added, in August of the same year,

'Phytologia Britanica
Witterings Botany in 2 vols.'

Of these sixteen books, seven—those marked with an asterisk—are still in the Society's Library, a remarkably low rate of wastage over two hundred years.

From the beginning, the Society had four specialist committees to organise and supervise its activities. They dealt with Agriculture and Planting, Manufactures and Commerce, Mechanics and the Useful Arts and Correspondence and Enquiry. They were elected annually, at the General Meeting in December and the main task of the first three was to decide on the premiums which should be offered and awarded. The first lists of these premiums, agreed in December, 1777, give a good idea of the improvements and inventions which at this time were felt to be particularly needed.

The list agreed on by the Agriculture and Planting Committee included seventeen items:

'1. Turnips (i.e. weight of crop)
2. Turnips and Beans (subsequently crossed out)
3. Destroying the fly on turnips
4. Hoeing turnips
5. Setting wheat
6. Winter vetches
7. Summer vetches
8. French or Buck wheat
9. Falling pigs[19]

10. Saintfoin
11. Carrots
12. Planting apple trees
13. Breeding and rearing calves for oxen
14. Rearing calves without milk
15. Onions
16. Planting birch
17. Planting bogs with ash'

The premiums which were considered to come within the province of the Committee on Manufactures and Commerce were:

'1. For Introducing the manufacture of Black Silk Lace
2. For working the said Lace
3. Increase of malt
4. Extract of oak bark for tanning
5. For tanning leather with the said extract
6. Ashes of Fern and Weeds
7. Rendering Hard Water Soft
8. Securing Boots and Shoes from Imbibing Wet
9. Marking Sheep without Pitch or Tar
10. Spinning Machines
11. For Spinning Yarn or Flax
12. For Spinning Yarn from Wooll
13. For Making Writing Paper without Rags'

The Mechanics and Arts Committee wished to encourage:
'1. Engine for Making Wooll Cards
2. For Winding Long Skained Silk
3. Machine for Conveying Green Winter Crops off wet Arable Land
4. Machine for sowing Carrot Seed
5.   Do.   Horse Beans
6.   Do.   cutting or bruising Woad
7. For a small portable Crane
8. Machine for destroying Vapours and producing Light in Coal Mines
9. Machine for floating pasture lands'

Each year brought new projects for premiums. In 1784, for instance, a long list of ideas and achievements which the Society was anxious to encourage included those 'to encourage industry in the children of cottagers', 'for raising crabstocks and grafting of the best kinds of cyder and chamber fruit', 'for reaping corn by women', 'for raising quickthorn for hedges by seed', 'essays on the art of making butter and cheese', 'for the best method of detecting the adulteration of medicines', and 'for ascertaining the constituent parts of a very rich soil and of a very poor soil by suitable experiments'. In 1786 a premium was offered for new types of packing paper, 'composed of vegetable substances not previously manufactured into cloth, thread or cordage and which shall be as

5. Report on varieties of Cider Apples, in Rack's hand, 1779.

cheap as similar kinds of paper now in use', and there was another 'for means of destroying smoke of fire-engines, glass-houses, furnaces etc. in order to prevent their being an annoyance to the neighbourhood'. A concern about the pollution of the atmosphere is certainly not as modern as most people suppose. Also in 1786 a premium of £10 for Friendly Societies,[20] offered for the first time, was awarded to the Friendly Society at Weston, 'consisting of 55 members, most of whom are Handicraftsmen and Labourers'. In the following year, 1787, the premiums included one for 'raising the greatest quantity of Honey or Wax from Bees without destroying the bees', a reminder that the techniques of bee-keeping were still very primitive. A large part—often as much as half—of the Society's annual income was spent on premiums, and the failure of members to pay their subscriptions could therefore be a serious matter.

A special category of award was to farm workers, for 'long and faithful servitude'. The first of these premiums was granted in December 1778 to Barnabas Marshall, of Enford, who had served Robert Baden for 24 years; Mary Hacker, of Puddimore, 30 years; and Mary Bennet, of Nettleton, 13 years. The first two received 3 guineas—the premium for a prize ram was 10 guineas—and the third 2 guineas. In 1779 a similar set of awards included one to Thomas Jefferies, of Wick, 'for 10 children born in wedlock and brought up without parish assistance', and in 1780 there was one to William Phillpot for '14 children, 7 now living, brought up without parish assistance'.

After a few years, the conditions for the award of 'premiums of long servitude' were revised, 'so as to have effect of exciting good behaviour in the younger class of servants', and a regular form of application was drawn up. The applicant had to have lived and worked in one place for five or more years and to declare his intention of continuing to do so for five years more. The form read:

'Sir,

I beg to inform the nobility and gentry of the Bath Agricultural Society that having lived as a hired yearly servant in the Station of .... with .... of this parish during ..... yrs ending the ... day of ...... last, I intend, if it please Divine Providence to grant me Life and Ability, so to live in Sobriety, Industry and Fidelity in the same place, as, at the end of 5 years from the expiration of my last, to claim a premium from the Bath Soc. with success. Witness my hand ....'

6. Form of application for Premiums of Long Servitude, c. 1785

In this first, golden period of the Society projects proliferated. Nothing was impossible, nothing too much trouble, everything was exciting. In February 1779, for instance, there was a proposal to list all the plants in Britain which were either eaten or rejected by cattle. The task was completed by August of the same year. Rack himself took a special interest in stone-breaking hammers and in improved wheelbarrows for use on the roads. Models of these were kept on the Society's premises for many years. To lessen Britain's dependence on American exports, it was proposed to grow tobacco in the South and West of England, but it was eventually decided that 'the cultivation of that plant does not appear to be an object which under the present Parliamentary restrictions can be of public utility'. An Improved Machine for Cooking with Charcoal did, however, materialise. Both the Manufactures and Commerce and the Mechanics and Arts Committees were impressed by the design submitted and recommended it 'for general use, not only in private families, but at the Camps, to which latter it seems peculiarly well adapted'. In August 1779, they sent a letter, recommending its use by the Army, 'to the Commanding Officer of the camp near Salisbury.' A specimen with an improvement for 'baking pyes, rolls etc.' was ordered for the Society and went on show at its premises.

Among the subjects which particularly engaged the attention of members and their friends during the 1770s and 1780s was rhubarb. This was Rheum Palmatum or True Rhubarb, the variety grown not for the sake of the stalk but of the root. It was imported from China and the Middle East and sold, dried and powdered, for use in medicine as an aperient. It was expensive and much-favoured by doctors. Consequently, there was every incentive to try to produce it in Britain. In September, 1778, Robert Davis, of Minehead, presented the Society with such a plant which he had raised from seed. It was tested and reported on in considerable detail by Dr. Falconer on the Society's behalf. He carried out experiments with patients at the General Hospital in Bath to discover its 'purgative virtue' and found that 'its operation was in every respect such as might be expected from the best foreign rhubarb.' The specimen was, in his opinion, 'extremely good in its kind; very little if at all inferior to the best brought from Russia or Turkey, and fully sufficient to supply the want of foreign Rhubarb'.

The Agricultural and Planting Committee, 'the cultivation of this plant being thought to be a matter of great consequence as an article of commerce', therefore handed over the specimens and report to the Committee of Manufactures and Commerce, which then became, temporarily, almost the Committee on Rhubarb. The rhubarb controversy brought a large correspondence and several more roots from different parts of the country. One member sent notes on rhubarb cultivation, given to him by 'a friend resident in Russia', and another

English Rhubarb No. I.

	Grains.	Stools.	
April 30	10	4	no griping.
May 1		3	
2	10	2	before 12 at noon with fome griping.
3		2	with fome griping.

Turkey Rhubarb.

May		Grains.	Stools.	
May 4		10	4	with confiderable griping.
5			2	with griping.
6			1	
7		10	3	with griping.
8			0	
				free from complaint.

2.   Mary Beft, aged 74; difeafe, violent pain in her ftomach, increafed by fwallowing, with head-ach and coftivenefs.

English Rhubarb No. II.

1785.			
Feb. 22		0	
23		1	
24	35	0	the powder taken at bed-time.
25		2	no griping.
26	35	0	the powder taken at bed-time.
27		4	flight griping.

7. Dr Falconer's rhubarb experiments on patients at the Royal Mineral Water Hospital, Bath, 1778.

gave instructions as to the best way of drying the roots. Dr. Fothergill wondered if the quality of rhubarb varied according to the soil on which it was grown and a Quaker doctor or pharmacist by the name of Sims advised Rack to be careful about wasting the Society's money.

The eighteenth-century diet, which was seriously lacking in fruit and vegetables, resulted in an obsession with constipation and, at a time when the range of medicines was very restricted, great reliance was placed on purging, which was frequently an agonising procedure, especially for those suffering from that not uncommon complaint, 'the stone'. In devoting so much time and energy to the cultivation and processing of rhubarb, the Society was not, therefore, engaging in a frivolous activity. If this much-respected but expensive aperient could be made more easily and cheaply available, the British people would certainly consider such a development a major benefaction, and the Society was well aware of this. Its letter-book, minutes and volumes of printed papers contain many pages of correspondence, articles and reports on the subject.[21] The end of the saga did not come until 1792. Robert Davis, of Minehead, wrote to complain that he had sent 10 lbs. of rhubarb root in 1784, 'as the quantity required to obtain the Society's then offered premium of £50'. He had expected to get this premium, but had apparently been disqualified, on the grounds that, in the same year, he had received a silver medal for growing rhubarb from the London Society. This medal, Davis insisted, was 'an accidental honorary token, procured for him by one of his friends and was not worth more than 5s', and so, in his disappointment, he had written to remonstrate with Rack. He received no answer to his letter, and therefore wrote again, 'Conceiving his application neglected', and insisting on having his 10 lbs. of rhubarb returned, or the £50 premium paid. Rack evidently replied—there is no copy of the letter—saying the rhubarb could not be returned, as it had already been used for experiments, but that he thought the Society would make him a present of a piece of plate equal to the value of the rhubarb. Five years later, after further pressure, he did indeed receive a piece of plate, valued at 5 guineas, 'as a full equivalent for the Rhubarb'.[22] The Secretary's job was no sinecure.

There were a number of similarly long-drawn-out negotiations which demanded tact and patience of a high order. One which demanded special care was the matter of the Society's medal. It had been decided that it would be a good plan to have a medal which could be presented from time to time to persons who had made an outstanding contribution to agricultural progress. Discussions about a suitable design took place in April 1779, and eventually one was chosen and sent to 'the principal die-sinker in Birmingham'. When a specimen of the medal was sent, it was found to be unacceptable, 'although ordered to be sunk in an elegant manner at 25 guineas'. The Society therefore returned it and there followed lengthy negotiations with the craftsman concerned, who felt that some payment was due to him. The Society then suggested that they should make 'some small present, as a compensation for his trouble, although from his failing in this undertaking he cannot in justice demand it'. In February, 1792, they sent him £10 and heard no more of the matter.

Three years after its foundation, the Society was in a position to begin what it had always regarded as one of its main tasks, the publication of reports, essays and correspondence which were accumulated in the course of its work. This, it believed, 'is the only method by which the various improvements and practical information suggested to them can be generally dispersed, even among those whom, from the nature of their institution, they are under particular obligation to serve'.

The first volume of *Letters and Papers* appeared in 1780.[23] Its 112 pages contained a variety of communications, sometimes anonymous, under such pen-names as 'a Gentleman Farmer in Norfolk', sometimes signed.

'A Gentleman near Norwich' mentioned a drill-plough, made by James Blancher, of Attleborough, Norfolk. An editorial note adds, 'We have one of these Drill-Ploughs at the Society's Rooms, with some new improvements made by the inventor since the above letter was written. It has been tried by our Agricultural Committee in a field, and found to deliver the grain with great exactness and regularity, quite to the satisfaction of the Gentlemen and Farmers, who attended the experiment. Any person disposed to have one, may be furnished with it, by applying to our Secretary, price $5\frac{1}{2}$ guineas, and carriage'.

The Society had been in the habit of sending a list of questions to the author of a communication, in order to obtain further information. The questions and answers were then printed together, as a sort of catechism. Here is an example concerning a grain-by-grain method of planting wheat.

8. Engraving of Mr Boswell's Norfolk Plough. From the *Universal Magazine*, 1783.

'Q 4th How many grains were dropped in a hole, and was the crop hoed?

A— Two grains were intended to be dropped, but this is often uncertain, from the unskilfulness or carelessness of the children who drop the corn. This crop was not hoed, which, although an excellent practice and much used here when wheat is sown broad-cast, does not appear so necessary when it is set.'

A comprehensive list of thirteen questions was sent by the Secretary on a printed form to every High Sheriff in the country, 'requesting him to procure answers thereto from some of the best farmers and send to the Society'. The Suffolk and Gloucestershire replies are printed here. They occupy eight pages.

John Billingsley contributes an 'Account of the Culture of Carrots; and thoughts on Burnbaiting on Mendip Hills'.[24] This was based on experiments carried out at his farm near Shepton Mallet at the end of the 1770s. He kept precise accounts from the time he ploughed the land until the crop was safely harvested. He communicated his costings to the Society, who published them in the hope that other farmers would be encouraged to follow his example. There were eight acres of carrots, and the figures were as follows:

		£	s.	d.
'February 15	First ploughing across the ridges of the cabbages, 45 per acre	1	12	0
March 1	First harrowing, 9d per ditto		6	0
April 15	Second ploughing, 45 per acre	1	12	0
April 20	Second (bush) harrowing, 9d per acre.		6	0
	30 lb. red Sandwich carrot-feed, 1s	1	10	0
24	Sowing by hand in drills, one foot apart, and covering the seed, 13s	5	4	0
June 4	Hand-hoeing and thinning, 20s	8	0	0
October	Digging up, 30s	12	0	0
	Carting home, cutting off tops and securing	10	0	0
	Rent of land	8	0	0
		48	10	0

The produce was 640 sacks of 4 bushels each, valued at 3s a sack	£96	0	0
Each sack weighed upwards of 200 pounds.			
Net profit of the crop.	£47	10	0

Or nearly £6 per acre. Quantity of carrots, 8 tons per acre.'

Carrots continued to attract the Society's attention for some time, evidence that they were a strange crop to farmers in Britain, or, at least in the South-West. The second volume of the *Bath Society's Papers*, published in 1783, included an article by Arthur Young, 'Proposal for further experiments on the Advantages of Cultivating Carrots'. This contained detailed costings per acre, from his own and other farmers' experiments.

All that men like Young and Billingsley were doing was to urge farmers to follow the example of successful merchants and tradesmen. Without proper records, a farmer had no real idea of how well or how badly he was doing and, as Billingsley puts it, 'no guide and signal-post in his future proceedings'. This insistence that farming was subject to the same commercial laws and procedures as any other business was perhaps the most important service which the agricultural societies performed during the eighteenth century. Accounting is an attitude of mind, a wish to impose order and method on one's affairs. Farmers who kept accurate records were more likely to appreciate the value of controlled experiments, and to keep an open mind towards new techniques. Unlike the men who were content to muddle through from one year to another, they were interested in efficiency, in whatever form it expressed itself.

Eighteenth-century farmers had no science to help them. Progress was based on an empirical approach and on trial and error. Communication was largely a matter of saying: 'I did this and I will tell you exactly what the result was'. Why something happened was another matter, and in many cases the answer would not be available for another hundred years or more. But, meanwhile, the Society carried out a valuable task by printing practical letters and articles, of which this is a typical example:

'In autumn, 1780,' wrote Joseph Wimpey, of North Bockhampton,[25] 'I went into the north of Devonshire, to spend a few weeks with a gentleman who cultivated his own estate. In October, when the cold rains came on, for many days running, he had one or more young sheep or lambs brought in, either dead, or in a dying state. They were one and all much swoln in the body, without any other visible difference from those that were well. In the field, I observed, they were much inclined to lie still, till rouzed and put up. I had several of them brought home and laid by the fire-side, and made several experiments upon them, but without success. At length, I had two or three of them opened, to see if I could discover any internal cause of their malady. All the viscera appeared to be sound and perfect, without the least sign of disorder; only, as I said, the body was greatly swoln.

'When the knife entered the belly, there flew out a great quantity of rarified air with a considerable noise, upon which the body immediately fell to its natural dimensions. As no unsoundness appeared in any of the viscera, I conceived the expansive vapour was probably the cause of the disorder, and the effect of obstructed perspiration, occasioned by the cold rains so common at that season. Upon enquiry, I found the disorder was Common in that county at that season of the year, and at that time was very rife for many miles around, which confirmed my suspicion as to the cause of the disorder.

'About sixty of these lambs remaining, I proposed to the hind (bailiff) to try if we could not preserve the remainder by sheltering them from the cold rains and damps of the night, by putting them into an airy barn, which was contiguous to the fields, to remain there till it might be thought proper to let them out in the morning. This effectually answered the purpose, for not one miscarried afterwards. In three or four days time, their coats, which appeared of a washed sickly white while they lay out, became a natural healthy-looking yellow, and they appeared as lively and healthy as at any time of the year. I should say therefore, if this method be pursued, many thousands may be preserved by its means.'

It was all very well for Dr. Fothergill to say that agriculture could not be understood without a knowledge of chemistry.[26] The truth was, at this date, the chemists were unable to answer the fundamental questions which were occurring to many progressive farmers. 'What,' asked Wimpey in another contribution to the Society's *Letters and Papers*[27] 'is that substance, matter or thing, which is the true and only proper food of plants; which enters into the vessels appointed by nature to receive it, is assimilated by, and become constituent parts of them, augmenting their magnitude, extension, and weight, from an almost imperceptible atom to the weight of many tons, and to a body of inconceivable dimensions?' Nobody could tell him, or, as he put it himself, 'Various are the opinions of the learned concerning this matter. Some suppose the food of plants to be water; some, earth; others air, nitrous salts, oil, etc. etc. perhaps all of them wide enough of the mark. It must be confessed, we know nothing of the essence of things. We are not endued with faculties equal to the curious research. Things are known to us by their properties only. But what are their properties by which they are known to us, but certain powers to affect us in a particular manner, and to impress different sensations and perceptions on our bodily organs? These different perceptions, indeed, enable us to distinguish, accurately enough, one thing from another; but we are totally ignorant of the nature of those powers, and equally so of the essence or substratum in which they inhere, and by which they are supported.'

But, without understanding the reasons for their success, experimentally-minded farmers often hit on a sound principle. Here, for instance, is a description[28] of a system of cooling meat immediately after slaughter, in order to improve its keeping qualities.

'Let them (the animals) be fasted a day or two in a cool house. Kill them in the evening, and as soon as the skin is taken off, hang the carcase between two doorways where there is a current of air. Then get a fan, such as is used for winnowing corn, and place it to windward of the carcase, and let a man turn the fan for the whole night. In the morning, the carcase will be cold and stiff, let the weather be ever so hot. A putrefaction will not immediately follow, because the fluids are at rest. Carcase butchers, and people that kill for the navy, would find their account in having slaughter-houses near to some rivulet of water, where a wheel might be placed to turn a fan and many carcases hung up at a time for the benefit of the wind.'

The Society felt that it might be useful to carry out experimental work on its own behalf, to supplement the information which it was collecting from members, and in 1780 the Secretary reported that they had:

'now taken a quantity of land into their hands, for the purpose of making experiments in Agriculture, under the immediate direction of a select Committee, who will keep an accurate daily register of their progress, expences and success, in each separate experiment; to be published in the next volume of their memoirs. This, they flatter themselves, will not only enable them to decide with certainty on many modes of practice, the propriety of which is at present doubtful, but excite numbers to join in support of an undertaking prosecuted with a spirit that must produce salutary effects to the community.'[29]

Ten acres were taken over at Weston, on the outskirts of Bath, on the farm of Mr. Bettel, one of the Society's members. Mr. Bettel carried out experiments on behalf of the Society, under the supervision of an Experimental Committee. The scheme continued for ten years and then petered out, mainly, it seems likely, because the division of management between the owner and the Society led to irritations and unsatisfactory direction.

All experiments cost money, and it is remarkable how much the Society managed to achieve during its early years on a very small income. The earliest statement of accounts to survive is for the year ending December 1783. This shows total funds in hand to be £477 1s. 6d., of which £277 1s. 6d. was at the bank and the remainder in cash elsewhere. Subscriptions so far received for the current year totalled £499 6s. 5½d., but many members were in arrears,[30] despite regular reminders. The problem appears to have been most serious among members who lived at a distance—in the Bath area the Secretary was usually able to apply a suitable degree of pressure—and in an attempt to improve the situation members had agreed to act as collectors in a number of the principal towns within the Society's area. By the early 1780s there were such agents in Bristol, Taunton, Wells, Shepton Mallet, Wotton-under-Edge, Devizes, Marlborough, Salisbury, Sherborne and, surprisingly, Puddletown—one would have expected Dorchester.

The biggest expense was always the payment of premiums, which absorbed half the annual income. The volume of *Letters and Papers* also took a large slice of the funds. Members received a free copy and non-members had to pay 5s., but the receipts from sales amounted to very little. As with all societies, there were members who expected the Secretary to achieve the impossible. A good example of this occurred at the General Meeting in December 1785, when a motion was proposed that there should be an annual dinner, which 'will tend to strengthen the general bond of union and give opportunity for such a free discussion of agricultural subjects as may prove of general service to the Institution'. This, it was generally agreed, would be a useful institution, since, with no Show or other function to bring members together occasionally, the Society, for a great many members, existed only on paper. But the officers of the Society felt obliged to resist the suggestion that the Society should pay for the dinner out of its own funds. For an annual subscription of a guinea a year, which was by no means always paid, a free dinner would have taken most of the balance at the bank, a fact which came as a surprise to some of those present at the meeting.

There were, however, certain very distinguished people who were excused their subscription, in exchange for the reputation which their membership brought to the Society. In February 1780, 'Monsieur Komhoff from Russia, now residing in England by order of the Empress for the purpose of studying agriculture' was elected an honorary member, as was Charles Blagden, M.D., F.R.S. In the previous year, Dr. William Watson, of London, and Dr. Ingen-Louz, of Vienna, were similarly honoured.

To have launched the Society and taken it through the enormous programme of work which was accomplished during its first ten years would have been a noteworthy achievement for a fit man, and Edmund Rack was a very sick man. He suffered seriously from asthmatic attacks, he was hard up—at some time soon after his arrival in Bath he lost his savings in some unspecified financial disaster—and he had to look urgently for other ways of adding to his income. One such opportunity was to write the parish-by-parish survey included in Collinson's great history of Somerset. 'This,' we read in Collinson, 'he indefatigably pursued during the successive years of 1782, 1783, 1784, 1785 and 1786, and, except a few towns and parishes, lived to finish.' And this at a time when, we are told, 'he could not, without the greatest difficulty of respiration, walk across a room; so that he rather existed than lived'. He died in February 1787, at his house at No. 5, St. James's Parade,[31] where the Society allowed him £30 a year for the use of two rooms as its offices and headquarters. He had moved there the previous year from a house in Harrington Place, Queen's Square, where he had been since 1781 and where the Society held its meetings, 'in a room commodiously fitted up for that purpose and for the reception of their books, models and machines.' The year before he died his salary was raised from £50 to £70 a year.

The Times gave him a four-line obituary, which does him much less than justice:

> 'On Thursday, died at Bath Mr. Edmund Rack, one of the people called Quakers. He was Secretary of the Bath Agricultural Society, and also the first mover in establishing it.'[32]

A more fitting tribute was paid to him a few years after his death, when a friend referred to the Bath Society as 'the unperishing memorial of his judgment, his benevolence, and his industry'.[33] His judgment was excellent—the time was perfectly ripe for setting up the Society—there is ample evidence of his industry—launching the Society, drumming up members and subscriptions, commissioning and editing articles, sending out and analysing questionnaires. There was also a suppressed practical man in him, given to inventing improved hammers and wheelbarrows. But, much more important than any of these things, was his benevolence. He really cared about his fellow men. One of the strongest and most moving pieces of writing he ever produced was his essay opposing the barbarities of the penal system,[34] and especially executions, 'those periodical exhibitions of human vengeance'. 'Men of humane dispositions and Christian principles, under all denominations, will ever consider the life of an offender to be of infinitely greater consequence than the loss of a horse, sheep, or a purse of money'. It was a courageous and dangerous attitude to adopt, at a time when property was supreme and rural England was ruled by prison and the rope.

Before he died, Rack could fairly claim that the Society had gone a long way towards carrying out the programme of activities which the founder-members had proposed. It had spent a great deal of money on premiums, to encourage innovation and experiment. It had published the kind of articles and letters which it believed would help to raise the general level of farming efficiency in the South-West of England, and it had built up a workable system of collecting useful information, from all parts of the country. Most important of all, perhaps, was its success in keeping its agricultural interests within the framework of the national interest as a whole. It showed an interest in prison reform, Friendly Societies, tithe-reform, rural housing and manufacturing, among many other topics, realising that progress had to happen on a wide front or not at all.

Inevitably, the Society attracted cranks, anxious to find a platform for their ideas, bores, and people with strange hobby-horses. All societies do. It also seems to have had its fair share of negative-minded men who saw snags and sinister possibilities in everything, men like Thomas Beevor who urged the utmost caution in awarding long-service premiums to maid-servants, since some of them had unmarried employers with whom they might well have been living for years 'in a state of carnality'. The Society could find itself, as a result, in the position of devoting members' subscriptions to subsidising immorality. The premiums continued, even so, to be awarded.

CHAPTER TWO

In Britain the 1780s and 1790s were characterised by two historical processes, one centred on the introduction of machinery into manufacturing and the other on the French Revolution. The first drew people together into towns and into larger and larger working units and caused great hardship among families who earned at least part of their living by carrying on handicrafts at home; the second brought about a boom in both farming and manufacturing, and once war with France had broken out, created a demand for munitions, ships and military and naval equipment of all kinds. It also led to a widespread fear, among the governing classes, that revolutionary ideas might cross the Channel. The repressive legislation which was aimed at preventing this served the secondary purpose of dealing with the protests and disaffection caused by industrial changes. The political atmosphere in Britain during the twenty years following the French Revolution was far from liberal. Habeas Corpus was suspended, meetings were made illegal, freedom of the Press was abolished and, for the first time, income tax was introduced, as a means of paying for the wars.

Wheat prices rose sharply, partly because the population was rising fast and partly because of the abnormal war need for provisions to feed British and Allied soldiers abroad. These high prices made wheat-growing very profitable, and caused great efforts to be made both to increase the yield and to extend the cultivated area. Meat prices also rose rapidly, and there was a demand for better breeds of cattle and sheep and for improved pastures on which to feed them. The war prices, more than anything else, were responsible for a remarkable speeding up of the enclosure movement, as it became realised that the traditional open field system was inefficient and stood in the way of technical progress in agriculture.

1787–1805: the next two secretaries

A Quaker, William Matthews was not without his troubles within the Society of Friends, where he was called to task for being too talkative. His predecessor, Edmund Rack, had also found himself in trouble with the Friends from time to time, especially over his habit of going on geological expeditions instead of attending Meetings. Yet, before he became Secretary of the Bath Society, Matthews had written several books on religious subjects,[1] and he obviously valued his Quaker connections. The reference to his un-Quakerly fondness for talking may be no more than an indication that he enjoyed the company of his fellow men, a quality which could have been valuable to someone who had to run such an oddly mixed body as an agricultural society, and to combine these duties with his other business interests.

His understanding with the Society must have been somewhat unusual, although there is no written record of it. From the beginning of his Secretaryship, he broke with the Rack tradition of working from his home and took the Society's headquarters back to Hetling House, where the original meeting had been held in 1777. He ran and advertised a seed and agricultural implement business from this address and announced that 'all orders from his Friends, in any part of the Kingdom, will be punctually answered'. In 1793, the *Universal British Directory* refers to him as 'Agent to Royal Exchange Fire Office', at Hetling House. He clearly had accommodating employers, but there is no evidence that he neglected his duties in any way.

One of his first acts was to draw up what had not previously existed, a complete alphabetical list of members, 'contrived to show the state of the subscriptions on imputation for six years to come'. This was a curious use of such a list, since it assumed that membership would remain at the same level for a number of years and that all members would pay their subscriptions, both of which conclusions were doubtful. However, this is the first list we have and it provides some interesting information. For the year 1787 there were 266 Ordinary and 55 Corresponding Members. The Ordinary Members were distributed over the country as follows:

Bath	46
Bristol	15
Buckinghamshire	1
Cornwall	2
Devon	5
Dorset	8
Gloucestershire	13

SEEDS, and SEED CORN,

OF THE BEST QUALITY,

And adapted for Change on different Soils,

MAY BE HAD, AT THE MOST REASONABLE RATES, BY PERSONAL OR WRITTEN APPLICATION, TO

WILLIAM MATTHEWS,

SECRETARY;

Who, in cafe of his refigning that Office, will continue the fame bufinefs at HETLING-HOUSE, as ufual.

He will alfo continue to fuperintend the Conftruction of, and caufe to be Manufactured for Sale, various

Valuable IMPLEMENTS and MACHINERY
used in HUSBANDRY.

☞ All orders from his Friends, in any part of the Kingdom, will be punctually anfwered.

9. Advertisement for William Matthews' seed and implement business at Hetling House, 1797.

Hampshire	1
Kent	2
London	11
Norfolk	3
Nottinghamshire	1
Oxfordshire	1
Shropshire	1
Somerset, other than Bath	58
South Wales	19
Warwickshire	1
Wiltshire	22
Worcestershire	1
Ireland	2
Residence not stated	53

It is interesting to note that Matthews himself was an Honorary Member of the Royal Agricultural Society of Lyons, France. A framed certificate confirming his election and dated November 30, 1787, is in the Society's possession.

The occupations of Members are not always given, but from the information provided certain trends are noticeable. There are a lot of doctors, a fair number of clergymen, and a sprinkling of other professional men. Members with military titles are not common and indeed, at this date, the list has a remarkably untitled look about it. The general impression is that of a middle rather than upper class Society, containing perhaps fifty really active[2] Members and no more than three or four downright and perpetual nuisances.

One very time-consuming member was Mr Henry Vagg, who saw possibilities of making a small fortune by persuading the Society that he was in possession of a secret invention, for which the agricultural community had long been waiting. The Society appears to have seen the red light just in time.

'The Bath Agricultural Society,' the Minutes record,[3] 'established on the principle of liberal communication, having thought proper not to give any sanction as a body to a late lucrative scheme of Mr Henry Vagg, to impart a Secret for the preservation of Turnips, on condition of a full subscription of £2,000:[4] and moreover as by a letter lately received by the Sec. from a dissatisfied subscriber to the said scheme, it appears that the writer expects the Interference of this Soc. in the said decisions. Resolved that it be proposed to ensuing General Mtg. 'that the Sec. be desired not to correspond officially on the subject with any person whatever and that he be desired on all necess. occasions to represent the Soc. as totally unconcerned in Mr Vagg's late proceedings.'

The wording of this Minute is diplomatic, but it is not altogether unreasonable to infer that the Secretary may have given Mr Vagg rather more encouragement than he should have done.

But, in the midst of such troubles and encumbrances, the serious business of the Society continued. The *Letters and Papers* appeared regularly, if not annually. Vol. IV was unfortunately late in appearing.[5] The delay in publication had several causes, 'among which, and not the least operative, was the decease of Mr Edmund Rack, the late useful and ingenious Secretary of this Society'. Another was 'the superintendance (sic) of the press, through the printing a second edition of the former three volumes—all of which are now completely reprinted'.

This volume had, as the Society always intended, a strongly regional emphasis. There was a report of a ploughing trial at Barracks Farm, near Bath, on 'a field of strong old ley ground', arranged to award a premium for 'the cheapest and best plough, for the common practice of husbandry in these parts of the kingdom'. There were six competitors. Billingsley won, with a double-coulter plough, drawn by six oxen. His man ploughed an acre in three hours four minutes. Dr Falconer contributed an article, 'On the Preservation of the Health of Persons employed in Agriculture', in which he listed their main troubles as rheumatism, colds, hernia, and the consequences of eating too much fruit and drinking beer and cider too fast. And, of interest both to dairy farmers and to the general public, an exposure of the effects of the near-monopoly which the London merchants had on the butter and cheese trade. 'I was once at Axminster,' said the author, 'where no bread and butter could be had with our teas; the reason being asked, the mistress of the inn assured us it frequently happened, that an ounce of butter was not to be got in town, unless on a market-day; for all the great dairies were under contract with the London dealers, for all they make, at a fixed price, which made it both scarce and dear. At the time she said this, there were 100 tubs of butter piled up in the gateway of the inn, in readiness for the London waggons.'

The centre-piece of Volume V, which was distributed to members in 1790, was a long article by Joseph Wimpey, one of the Society's most assiduous supporters, 'On the Improvements in Agriculture that have been successfully introduced into this Kingdom within the last Fifty Years'. Wimpey instances particularly greater attention to tillage; much improved ploughs; a double-mouldboard plough for earthing up potatoes; drilling or setting grain instead of broadcasting it; rotation of crops; more, and more intelligent, manuring.

New crops had been 'transferred from the garden to the field. Turnips, potatoes, cabbage of different kinds, carrots, parsnips, etc. were cultivated for domestick uses, long before the time proposed; but the field culture of these articles for the feed of cattle in any considerable degree, is quite a modern practice. The success which hath attended the use of these articles, hath incontestibly established their great value and importance; but unfortunately their culture hath hitherto been much confined, and is very far from being generally practised.'

Sainfoin and lucerne, he observed, were now more widely grown, and

'Another new article which has been very lately introduced is the *Mangel-Wurzel*, or Scarcity Plant. From the success some few gentlemen have had in its cultivation, it seems to promise to be of the greatest utility for the feed of cattle. However, it is very little known as yet, it being supposed that not one farmer of a thousand has so much as ever heard of the name. It is generally agreed to be a species of the beet, of which there are many. The seeds of both have exactly the same appearance, and the leaves and roots differ only in colour and size, for the manner of their growth is exactly the same; but the leafage of the new sort is said to be much more luxuriant and abundant, and the roots vastly larger.'

There was a report of a 'Publick Trial of Ploughs and Drills', near Devizes, in April, 1790. It was organised by the Society's 'Committee of Gentlemen Farmers'. 'It was expected that six ploughs of different descriptions would have started for the premiums of the Society, but only four were found in the contest; one or two haveing declined, on account of the difficulty of the work, and a new swing plough, lately invented by the Rev James Cooke, being delayed on the road by the carrier.'

Dr Fothergill had been busy with his investigations once again, this time in connection with metallic poisoning. Vol. V contained two articles from him on the subject, the first 'Of the Poison of Lead, with cautions to the Heads of Families concerning the various unsuspected Means by which that insidious Enemy may find Admission into the Human Body', and the second, 'On the Poison of Copper'.

'Is it not a moving spectacle,' he asked, 'to see poor industrious tradesmen, particularly the manufacturers of red and white lead, daily exposed to the noxious fumes of this pernicious metal? Unhappy men, whose hard lot it is to earn, by the sweat of their brow, a scanty maintenance, breathing all the while a tainted air, and inhaling a slow poison at every pore, in order to prolong a wretched existence! These surely have a peculiar claim to our compassionate regards, and I should think myself happy if I could awaken the attention of this Society towards the alleviation of their sufferings.'

Wine, beer, cider, vinegar and other liquids commonly had lead added to them, to reduce their acidity. In France, he noted with approval, this had been made a capital offence. Other countries, Britain included, should follow the French example. 'Deliberately thus to adulterate the common articles of life with a slow poison, and wantonly to sacrifice the lives of innocent persons to unfeeling avarice, seems a refinement in villainy at which human nature revolts, and which could hardly be credited in a Christian country! The savage tribes of the most barbarous nations, who attack their declared enemies with poisoned arrows, are never known to discharge them at their unoffending neighbours and countrymen.'

No section of the community was safe from such practices. Danger lurked even in the nursery. 'Children's Play-things are commonly painted with a com-position of read or white lead; but how often do we see the smiling innocents suck within their lips those pernicious toys, while the unsuspecting parents look on with apparent satisfaction? The application of an ointment with litharge, or white lead, to nurses' sore nipples, often proves fatal to sucking infants.'

And the vanity of women made them particularly vulnerable to these manufacturing malpractices. 'Were it permitted us to penetrate the secret recesses of the toilet, and to explore at leisure the nature of the mysterious articles which administer so conspicuously to artificial beauty, we should probably find that some of the most celebrated cosmeticks consist of preparations of lead, mer-cury, or bismuth.'

Copper, in his opinion, was as dangerous as lead, especially in the preparation of food and drink. 'Notwithstanding every remonstrance to the contrary, copper and its compounds continue to disgrace not only our kitchens, dairies and con-fectionaries, but also the breweries, distilleries, laboratories, and even shops of the apothecaries.' Only iron, glass, and pottery could be considered really safe.

In Vol. VI (1792), the Society's Secretary himself felt drawn to philosophise at some length on the current state of agriculture. On the whole, he believed, things were going pretty well. 'The landed gentlemen, and those daily enriched by com-merce, are now emulous in the study of agriculture; the improvement of poor, waste and barren lands, is become a favourite undertaking; and a laudable enquiry seems to be general, How the face of the country, according to its local circumstances, can be rendered most productive? This general enquiry, and the consequent exertions, may in no small degree be imputed to the publick-spirited institutions in the kingdom, among which THE BATH AND WEST OF ENGLAND SOCIETY has the honour of holding no inconsiderable place. The effect of such establishments, though gradual and diffused, has undoubtedly been sure, and happy. For under all the circumstances of the increased demand, it is an obvious truth that the supply of every necessary, and most of the comforts of life, is not only abundant, but in general easy of acquisition to the honest and in-dustrious of all descriptions. This augmented supply has been furnished, in some degree, by the increase of lands brought into cultivation; but perhaps far more by improvements in the general system.'

New premiums were announced, among them two of great political impor-tance in the South-West, the first, 'for ascertaining in the Western Counties, by any experimental method, the best breed of sheep in Carcase and Wool', and the second, 'for the best practical Essay, founded on experience in raising Apple-Stocks, and the most successful method of grafting and raising apple-trees for the orchard; together with the best essay on gathering apples, making them into cyder, and of managing that cyder until it shall become fit for use.'

Circular letters were sent to members, in order to obtain information on par-

ticular subjects. One of these had asked for 'Observations on the supposed
Neglect and Scarcity of Oak Timber'. This produced a detailed reply from
Thomas Davis, the steward on the Marquis of Bath's Longleat estate, which the
Society wisely decided to print in the 1792 volume of its *Letters and Papers*.
There was, he believed, no general shortage of oak for ship-building. 'Luckily,' he
said, 'there are thousands of acres in this kingdom where *oak* is *the weed of the
country*', and improved transport was making it easier to use. 'A great deal of
oak in distant parts of the kingdom,' he was sure, 'will now find its way to the sea-
ports, by means of the many canals in the kingdom, which formerly were consumed only in the domestick uses of the county where it grew, while those same
canals will bring back deal at a cheaper price to supply those domestick uses.'

Once builders were able to obtain sufficient quantities of good quality
softwood, mainly from the Baltic, they turned away from the traditional oak.
'The uses of oak,' wrote Davis, 'lessen every day. Houses were formerly built
almost entirely with oak timber; but now the innumerable new houses in Bath,
Bristol, London, Manchester, Birmingham, etc. have very little oak in them. Deal
answers the purpose at a much cheaper rate.' And deal did not necessarily have
to be imported. 'The English grown spruce and silver fir timber are fully equal to
any white deal we get from abroad. The Marquis of Bath has used English grown
fir, for almost all domestick purposes, in the dry, for 20 years past, and finds no
wood except oak equal to it; and we have an instance of a cart-house, which has
been built with English grown fir upwards of 70 years, now almost as perfect as
when new.'

On the other hand, there was an enormous and increasing demand for oak-
bark for tanning and for oak-timber for beer-casks, for wheel-spokes and for
laths. Coopers and wheelwrights were interested only in straight timber,
however. They had no use for the crooked and curved pieces, which were the
very parts of the tree in particular demand for ship-building. All in all, Davis,
believed, the situation was well in hand.

Volume VI was almost a timber and woodlands issue. In addition to Davis's
contribution, there were a number of observations on the damage caused to trees
by squirrels and a series of letters on the management of woods. These included a
letter from John Ward, Lord Ailesbury's steward, mentioning that Ailesbury
'has introduced a covenant in his leases, whereby his tenants engage to *plant and
preserve a certain number of trees yearly*, in proportion to the size of their estates; but even this is not fully complied with, and he has lately employed a person to go over his farms, to seek out the fittest places in hedgerows etc. for planting and afterwards sent plants from his own nurseries and had them planted.
The same person marks for reserves any self-planted trees he can find in the cop-
pices and hedgerows that come in course for cutting to save them from being cut
down with the underwood.'

In general, Lord Ailesbury found his tenants ignorant and philistine about trees and 'as backward in raising timber as they are industrious in pollarding what does get up'. As a tree enthusiast he would have been fully sympathetic to another article in the same volume, the theme of which was the need to take advantage of the opportunity provided by enclosures in order to increase the tree population of the countryside. The author made a plea for making hedges of mixed trees, including chestnut, and for planting fruit trees in them.

But, however indifferent or hostile tenant farmers may have been towards trees—trees, after all, brought them little in the way of income—many landowners displayed a passionate interest and the Society did its best to keep them up-to-date with the most progressive thinking on the subject. Thomas Davis's survey of the oak situation has already been mentioned, and the 1795 volume contained another important essay by him, this time on the management of woodlands.[6]

Davis was one of the Society's star performers and most active members. He wrote well and he had a gift for marshalling a large number of facts in a lucid and convincing way. But, like others who contributed to *Letters and Papers*, he was much helped by what one might describe, in a modern phrase, as the Society's—or perhaps one should say the Secretary's—journalistic sense. In journalism, one gets what one goes to find out and the Society's choice of subjects for premiums and questionnaires showed, in general, a sure touch. The *Letters and Papers* dealt with live issues.

One is struck again and again by the excellent balance of each volume. Technical and practical articles are judiciously mixed with others with a social or political emphasis, and the refreshing eighteenth century habit of calling a spade a spade gives an attractive flavour to these old volumes and makes them very good reading even today. It is enjoyable to find Mordaunt Martin expressing a low opinion of the people he employed to hoe his mangolds, 'women who ran home for every shower, and staid there if they saw a cloud',[7] and Dr Fothergill's belief that tea-drinking led women on to alcoholism is expressed in an equally forthright fashion. 'This relaxing beverage,' he insisted, 'poured down hot, as it generally is, at least twice a day, tends to unnerve the female frame and produce universal languor. The natural spirits being depressed, recourse is imprudently had to artificial ones, the property of which is, first, to wind up the springs of the animal machine far above their natural pitch, then suddenly to let them down as far below it: hence it is that each glass of spirits soon requires two more to obviate its own bad effects, and the remedy at length is discovered to be ten-fold worse than the disease.'[8]

The printed volumes of *Letters and Papers* contain no clue to the dissatisfaction which the Secretary felt with the conditions under which he was expected to work. The Society's activities multiplied, its prestige grew, its publications appeared and the outside world would have had no reason to suppose that Hetling House was a centre of anything but harmony and contentment. In November 1790, however, Matthews had threatened to resign, unless his life was made more tolerable.

'When I first had the honour of being appointed Secretary,' he wrote to the President,[9] 'I felt a suitable gratitude for the almost unanimous Suffrages of the Society. I resolved always to consider the Emoluments of office as a secondary Object, and to give Information whenever I should find the necessary Confinement incompatible with my other Duties. I now have to acquaint the Society that a period is arrived, when from various circumstances I shall find it necessary either wholly to relinquish my Trust, or devote more of a Handsome Salary than can well be afforded to the payment of a capable Assistant, who may be constantly on the Spot in my Absence: and as I suppose the Society may not find it difficult to appoint another Secretary of Abilities equal if not superior to mine, I beg leave to propose resigning at any time which shall prove most convenient to the Soc; holding myself bound and shall cheerfully undertake to discharge the Duties of my Trust till an Election can be made to the entire Satisfaction of the Soc.'

In the event, Matthews agreed to remain for another year, on the understanding that the Committee should make clear the hours when the Secretary could be expected to be found in his office, at the service of members. The official record reads:

'In order to obviate some Difficulties about the official confinement and proper liberty of the Sec (and the more to induce the present Sec. to continue in his office) it shall be considered henceforward that the Sec. of this Soc. endeavouring to do his duty faithfully, as well abroad as at Home, shall only be expected to give attendance at the Socy's Rooms between the hours of 11 and 3 o'clock, and that at other Hours and in cases of necessary Absence, it shall be deemed sufficient that an intelligent person attends on his Behalf.'

The arrangement appears to have met the requirements of both parties, since Matthews remained, not for one more year, but for ten. These ten years were an immensely productive period in the Society's history, not least because first the French and then the American war, together with the need to provide food for a rapidly rising population, kept agricultural prices and profits high.

Stockbreeding began to occupy an increasing amount of the Society's attention. A 'publick exhibition of fat Sheep of different lands' was held in the courtyard of Hetling House on the occasion of the Annual Meeting in December 1790, and this, the origin of the Annual Show, was repeated for many years. More important, from a scientific point of view, was the opportunity to acquire some of the Spanish sheep which had been imported on behalf of the King, for breeding experiments supervised by Sir Joseph Banks. The fine, long-stapled wool from these sheep was greatly prized by the cloth manufacturers and, with the woollen industry so important in the West Country, the area served by the Bath & West Society was an obvious choice for further attempts to build up flocks of pure or cross-bred Spanish sheep.

A lengthy correspondence, helped by a certain amount of personal influence, eventually succeeded in persuading Sir Joseph to make a small selection of these rare animals available to the Society. In August 1792, he wrote to say that he had received the King's command 'to select *two rams* as a present, the one a young sheep bred by his Majesty, at Windsor, of pure Spanish blood. The other an older Sheep imported from Spain in 1790, and used in his Majesty's flock last year. The older sheep is marked on the horn No. 8 and younger No. 26. I should not have selected so old a sheep for the Bath Society but under an Idea that they would chuse to possess an original one bred in Spain. And tho' No. 8 is an Old Sheep I have no doubt that he will be able to do his Duty, this year at least and possibly the next.'

The two rams were duly collected from Windsor by the Society and allocated, one to Lord Ailesbury, at Tottenham Park near Marlborough, and the other to Billingsley, 'for benefit of members'. No complaint was received from Lord Ailesbury, but Billingsley was far from satisfied. A month after No. 8 had been left in his charge, he wrote indignantly to say that Sir Joseph Banks' gift 'appears, from age and infirmities, to be totally incapable of serving the ewes intended to be sent to him.' The Society thereupon decided to write off No. 8 as hopeless, at least for the time being, and to pin their faith on No. 26, who was to be supplied with 80 ewes, 4 each from 20 members, the selection being by ballot. If the old ram showed any signs of improving as a result of Billingsley's care and encouragement, ewes were to be taken to him as well. Members whose ewes bred successfully from either of the rams were asked to keep detailed records of the progeny.

There were, however, certain consequences which the Society could hardly have foreseen. In November 1792, Lord Ailesbury's bailiff wrote to complain that 8 ewes had been sent by some totally unauthorised person, and to announce that he intended keeping them until he was furnished with an explanation. The Society's correspondence and minute books unfortunately contain no information to indicate how the matter was settled, but it is evident that breeding went on for several years. The old ram was dead by April 1795, and Lord Ailesbury was asked to report the sad fact to the King. By July the Society had received a letter from Sir Joseph Banks, offering a second gift from the King of two more.[10]

Since the whole point of the experiments with the Spanish rams was to assess the value of the resulting cross-bred fleeces to the cloth manufacturer, it was important to find a clothier who would co-operate in the necessary research. Such a man was discovered, Mr Joyce of Freshford, and he received a bounty of five guineas for his trouble. Unfortunately, fleeces from the Spanish ram cross which had been sent for washing were mixed up accidentally with others, which significantly reduced the value of the investigation. Mr Joyce continued even so with the undertaking and the Minutes for 1798 contain a full record of the results. The clothier had to provide accurate costings for 'sorting; dyeing and securing; drying and beating; picking; scribbling; oil; carding; spinning; list; glue; warping; weaving; burling; braying; milling; soap; dressing and drawing, as severally paid for to his ken, under the disadvantage of this small manufac[c].'

It is necessary to remember that, at the time when West Country farmers were busily crossing Spanish rams with their own sheep, knowledge of the principles of stock-breeding were of the most rudimentary kind. A paper by James Anderson, F.R.S., in the 1796 volume of *Letters and Papers* set out the most advanced thinking on the subject. It is a masterpiece of lucid exposition, with a beauty of style which scientists have long since lost. There were, says Anderson, certain remarkable specimens which occurred from time to time, 'different from others, though they still possess the general characteristicks of the parent breed. And so strong is the propensity of nature in all cases to produce its own kind, that if the individuals possessing these qualities, thus, as we would say, accidentally

10. Research by Mr Joyce of Freshford into the value of crossbred fleeces to the cloth manufacturer. From *Minutes*, 1798.

produced, whether beneficial or hurtful, be selected and put to breed with others, that possess qualities somewhat of the same sort, it is found that the descendants of these selected animals will, in general, be possessed of the distinguishing peculiarity for which they were selected in an eminent degree; though among these also some individuals will be found to have less of it than others. And if these least approved individuals be banished from the selected stock; and those, both males and females, which possess the wished-for quality in the most eminent degree, be put to breed together, the descendants of these will be still more improved; and by continuing this mode of selection for a great length of time, the improvement, as to this particular quality, may be carried to an indefinite height. In this way may be produced an improved breed; which, though agreeing in the general characteristicks with the parent stock from which it was selected, may possess some peculiar qualities in a much higher degree than it does.'

Nowadays, all this seems very obvious, but in the eighteenth century it was not at all obvious, and the Society was performing a service of the first importance to the farming community by helping to make such ideas better known and by encouraging carefully planned experiments to test them out. By 1802, it was known, for example, that wool from sheep not recrossed deteriorated and that constant interbreeding with Spanish rams was necessary to maintain quality.

Throughout the Society's records, published or in manuscript, between 1790 and 1810 there is a paradoxical and saddening contrast between the prosperity and enterprise of the farmers and the wretched condition of the labourers and their families. As Thomas Davis pointed out, in a very long article on the agriculture of Wiltshire, published in Vol. VII of the *Letters and Papers* (1795), farm families had had an important source of income removed when the cloth manufacturers began to spin wool by machinery in their factories, instead of having it done by women and children in the villages. A further cause of rural poverty had been the loss of free fuel, as fields were cleared and ploughed to grow the wheat that was so profitable at wartime prices. A temporary solution to this, Davis felt, might be for each parish to keep a few acres of furze unploughed. 'This,' he thought, 'might be sold for fuel, to those who could afford to buy, and given, instead of parish relief, to those who could not. Those who have hearts to feel for the distresses of the poor would, by this expedient, gratify their humanity; and those (if there are any such) who feel only for the preservation of the hedges, would find this a more effectual way to prevent wood-stealing, than a whip or a prison.'

However hard they worked, farm labourers could no longer 'live by their industry'. The parish had to support them from its poor rate; and, even then, the poor were forced to steal, in order to feed themselves adequately. In February 1798, the Society's Minutes refer to 'frequent depredations on Turnip Fields', and in the following year it adopted a resolution asking Parliament for stronger penalties for this offence, having 'received many complaints from its members of the frequent and daring Depredation committed on the growing exposed Produce of their Fields, now become much more valuable from the improved state of Agriculture, such as Turnips, Cabbages and etc.' British farmers were now, in other words, growing crops which were worth stealing. For some, the remedies were flogging, hanging and imprisonment, but for others the solution lay in providing farm families with decent houses and good gardens, where they could grow for themselves the vegetables they were now compelled to steal.

But, as Matthews himself pointed out in his introduction to Vol. VIII (1796) of the *Letters and Papers*, rural misery and rural crime constituted a circular process. The worse conditions were for the farm labourer, the more certain it was that only the lowest types would remain in the countryside. 'To remedy these evils,' he said, 'it is to be lamented that country gentlemen, and other considerable land-owners, are so little attentive to rural policy in the improvement of cottages, and the annexation of small pieces of land, for orchards and gardens, thereby to allure and fix the most active and useful of the peasantry.' A progressive house-building policy would ensure that 'the old country cottages and miserable huts, in which indolence, dejection, disease, and indelicacy, have been long propagated, will gradually become improved and re-built; and the allotment of land for useful garden purposes will become increased, to the improvement of the inhabitants in the essential articles of industry, health, decency, order, and contentment! The country would thus by degrees, and perhaps not by slow ones neither, acquire a new face of *civilisation*, respectability, and ornament.'

If farm-workers were given an opportunity to live decently, and to take a pride in their homes, their work would benefit, 'whether the business to be done be the cleaning of a stable, a pen, or a fold for cattle; of a farm-yard, a pond in the field, the making or mending of a ditch, the planting or plashing of a hedge, the grubbing up of weeds or brambles, the mending of a road, or whatever else in these common offices of the labourer; any or all of them will be done the better, by how much the labourer has been accustomed to value conveniencies, and the appearance of neatness in and about his own dwelling.'

The Society regarded the rehousing of farm workers as a matter which deserved a very high priority. This was emphasised by a new premium, offered for the first time in 1801. It was for the landowner who could show that he had built 'the greatest number of cheap durable and comfortable cottages, in proportion to the extent of his estate, for poor industrious Labourers to inhabit and who shall annex a portion of land not less than a $\frac{1}{4}$ of an acre to each cottage.' An earlier premium (1791) had similar aims, but was less comprehensive. It was for a design for 6 cottages, 'the most roomy, healthy and conveniently divided, containing not less than 2 Rooms a Floor and wch. shall not exceed the cost of £40 or at most £50 each cot., saving in price to be a principal Merit'.

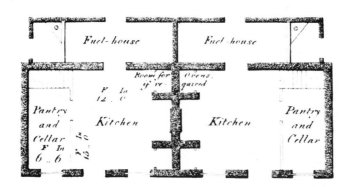

Fuel-house   Fuel-house

Room for Ovens, if required.

F In
12 . 0

Pantry   Kitchen   Kitchen   Pantry
and   and
Cellar   Cellar
F In
6 . 6

Chamber   Chamber   Chamber   Chamber

Scale of Feet.

9 8 7 6 5 4 3 2 1 0   10   20   30

11. Plan of 'Large Double Cottage with Skillings Behind'. Bath Society's *Papers*, Vol. VII, 1795.

Building new cottages may have been advocated, at least partly, because it was considered to be sound business, but the Society's concern with abolishing the practice of using small children to sweep chimneys was purely humanitarian. In 1795 there was a new premium which represented a serious attempt to deal with the problem. 'As the present method of sweeping chimnies is productive of great hardships,' the specification began, 'and sometimes of fatal effects, to the children employed, a reward adequate to its merits will be given for any Model which will effectually perform that Business'.

Unfortunately no effective Model was forthcoming, and in 1802 the Society tried again, with a more forcefully worded description of what was required. A premium of 6 guineas would be paid to a 'person who on or before Nov. Mtg. 1803 shall invent and fully describe a simple and efficacious method of sweeping Chimneys that shall prove satisfactory to the Soc. without having recourse to the barborous means now in use of forcing Young Boys to ascend, and carry on that dangerous and disgusting Business Employment so productive of distortion of the Limbs and other irremediable diseases'.[11]

It is possible that the fact that Edmund Rack and William Matthews were both Quakers may have had something to do with the strong social conscience which the Bath & West Society, as a body, showed for its formative first twenty-five years, when traditions and attitudes were established. Its interests were by no means purely technical and, indeed, its humanitarian principles tended to be somewhat more reliable than its judgement in technical matters. In 1794, for instance, it decided that it 'cannot venture to give its Publick Recommendation' to the principle of ball-bearings, patented by Mr Morgan of Carmarthen. Having inspected a model, the Society had to admit to 'strong doubts existing in the minds of some Gentlemen as to the Durability and Service of this Movement in heavy carriages'. Fifty years later, the decision would almost certainly have been different.

More serious, however, was the rejection of 'Mr Griffiths' machine for cutting down corn in the field'. This, it was felt, showed ingenuity misapplied. 'The Committee has doubts,' Mr Griffiths was told, 'whether any mechanical invention of a complicated Nature can be applied to the cutting down of Grass and Corn that will supercede the single scythe, naked or cradled, and used in the common way ... a way easily learned, effectual to its end and affording useful and agreeable exercise to a hardy peasantry.' This is one of the classic instances of poor judgement on the part of those who had the power to advance or kill a new development. If the Society's Committee had come to a different conclusion, a British reaping machine might have come forty years earlier than McCormick's invention in the United States, where there was no feeling at all that mowing was useful and agreeable exercise for a hardy peasantry.

But everyone has the privilege of making mistakes and, year by year, under

Matthews' guidance the Society continued to look for new solutions to old problems; improved vellum from sheepskins, a recipe for killing slugs, flexible tubes for getting fresh air into coal-mines, better varieties of apples. Attendances at Annual Meetings were good—130 in 1797—the Dinners at the White Hart were highly thought of, 'a publick Exhibition and Sale for choice samples of Sheep and Cattle on the day of the A.M. of this Society' in Beaufort Square became a popular local institution.[12] The first legacy to the Society came in 1796, from Mr Benson Earle, of Salisbury and, to commemorate the occasion, it was resolved 'that the Bequest in Mr Earle's will be copied as a Tablet in Gold Letters and hung up in the Society's room.' £204 2s 8d was received, in June 1797, and invested in 3% Consols, the first step of this kind which the Society had undertaken.

In 1799, with his retirement from the Secretaryship approaching, Matthews felt entitled to take stock of what the Society had achieved. He did this in the form of a long introduction to Vol. IX of the *Letters and Papers*. The general theme of his essay was the necessity for greater agricultural production, in order to sustain a growing population. 'A slovenly farmer,' he wrote, 'is one of the greatest internal enemies of his country. To a man of common discernment, the appearance of many of our lands, in particular districts, is at once mortifying and disgustful; to the benevolent and patriotick observer, it is often *affictive*. For though the general fertility of this comparatively happy island, under the benign influences of the Author of Nature, is such, that *famine* rarely invades, however unworthy of universal plenty, the mass of its inhabitants; yet it cannot but be known, that pining *scarcity* is the frequent portion of some, against whom no particular indolence or moral turpitude may be imputable.'

Matthews was not a young man and he may well have felt inclined to devote his remaining years to his private business affairs, free from the restrictions and irritations which the Secretaryship inevitably involved, especially as the Society became better known and more influential. There had been an example of what might be termed Secretary's frustration in 1793, when the Society was involved in a disagreeable controversy over tithes. A letter, from John Franklin, of Llanmihangle in Glamorgan, printed in Vol. VI of the *Letters and Papers*, brought strong objection from the Revd Francis Randolph, Curate of Corston, and preacher at the Octagon Chapel in Bath, who regarded the letter as insulting to the clergy. The Committee of Correspondence and Enquiry looked into the matter, decided the Editor, Matthews, had been at fault and made a recommendation aimed at preventing such an unfortunate happening from occurring again. 'Whereas through inadvertance,' the Minute reads, 'reflections of a disagreeable Nature have appeared in the papers of this Society, to prevent such consequences in future, and with a view of more fully promoting the object of this Society, it is directed that in future the papers for publication be carefully examined by its Committee of Correspondence and Enquiry, and that every paper deemed proper for insertion in the Soc's Volume be signed by 3 Members'.

This amounted to a censure of the Secretary and Editor, and Matthews, who, as a Quaker, had no particular reason to wish to keep on the right side of the established church at all costs, must have been suitably annoyed. However, he soldiered on until January 1800, when members were officially told of his wish to resign and given notice of a meeting to be held in February to consider the appointment of a new Secretary. In his letter of resignation, Matthews referred to 'personal considerations of much importance to myself disposing me to resign, though with reluctance', without giving any indication as to what these 'personal considerations' might be.

At the February meeting, Mr Nehemiah Bartley, a nurseryman, of Lawrence Hill, Bristol, was appointed by ballot to serve until the next Annual Meeting, when the matter would be considered again. At the Annual Meeting, Bartley was confirmed as Secretary, and Matthews, who clearly retained the confidence of the Society, was made a member of all the Standing Committees. The Society was much indebted to him, not only for the immense amount of work he had done in the course of his Secretaryship, but also for providing a suitable headquarters. At some time during the 1790s—it has not been possible to establish the precise date—he had bought Hetling House and thereafter he was paid a rent for the rooms occupied by the Society. After his resignation, he continued to run his business affairs from the same address, although he 'exchanged rooms' with the new Secretary, in order to forestall possible criticism that he had not really given up control.

Bartley, it is evident, was a nurseryman on a considerable scale. His 1799 catalogue survives among the Society's collection of printed pamphlets. It has 35 pages and compares very favourably with the other eighteenth century nurserymen's lists which are preserved in the Library of the Royal Horticultural Society. There are, for instance, 124 varieties of apples, 48 of gooseberries and 13 of apricots. In addition to trees and shrubs, there were 'Kitchen Garden, Grass and Flower Seeds of all Sorts; Hothouse and Greenhouse Plants; Matts and every Article in the Nursery and Seed Line'.

He had farming interests as well, and won three of the Society's premiums. In 1787, he received 'for his account of experiments in Husbandry, and a Course of Crops, a Piece of Plate, value £3.3.0'. In 1788 he had 5 guineas' worth of plate, 'for the best crop of Buck Wheat', and in 1803 the Society gave 'Mr Bartley, the Secretary, for four fine clothing-woolled shearling Rams, £5.5.0'. Members were entitled to believe that they had chosen a man well experienced in agricultural matters to take over from Matthews. Unfortunately, however, his administrative ability did not turn out to be of a high order.

Matthews was a meticulous man, and took great pains to make sure that

At the NURSERY PLANTATIONS,

LATE DAVIS'S,

LAWRENCE-HILL, near BRISTOL,

ON THE UPPER TURNPIKE-ROAD TO BATH;

NEHEMIAH BARTLEY,

Honorary Member of the Bath and Weſt of England Society,

Sells, on the most moderate Terms,

All SORTS of FRUIT and FOREST TREES,

VIZ.

Peach, Nectarine, Apricot, Apple, Cherry, Plum, Pear, Raſberry, Gooſeberry, Currant, &c. Oak, Elm, Aſh, Beech, Hornbeam, Black and White Poplar, Maple, Birch, Larch, &c.

Engliſh and Foreign Evergreens.—Variety of Ornamental Shrubs, &c.——White-Thorn Plants of different growth; Crab, Cherry, and other Seedlings.

Has a large quantity of remarkably fine, trained Standard Peach, Nectarine, Apricot, Cherry, Plum, Pear, &c.

GRASS AND GARDEN SEEDS.

N. B. Catalogues delivered gratis at the Plantations, and at the Bath and Weſt of England Society's Rooms, Bath.

12. Advertisement for Nehemiah Bartley's nursery at Lawrence Hill, Bristol. *Letters and Papers*, 1797.

everything was handed over in proper fashion. In April 1800, he produced a complete inventory of the Society's property, 'as delivered up consequent on his resignation', and in the previous month a strongly worded circular letter was sent out, in an attempt to recover some of the arrears of subscriptions. This reinforced a more tactful campaign which had been carried through the previous year, when the tone of each letter was brilliantly adapted to the recipient. A Mr Adams, for example, was informed that the records showed 'a chasm from April '87 to Nov. '89'. Matthews added, with consummate politeness, 'I have filled up the chasm by taking it for granted that I am wrong'. The Duchess of Newcastle merited a different approach. The Duke, it was noted with regret, had recently died, and out of delicacy no approach had been made to her earlier. The Society applied to her now, however, 'as she would not wish such a name to appear in arrears'.

One of the first major tasks to be undertaken during the new Secretary's term of office was the complete revision of the Premium Book. This was carried out during 1801 and the new list divided the premiums offered into 10 groups, 'Agricultural Operations; Soils and Manures; Crops and Plantations; Live Stock; Wool; Mechanicks; Chemistry; Useful Arts; Industry and Good Behaviour; Essays.' At the same time, a recommendation was made that a permanent committee should be set up to keep the premiums under review and to issue a revised list every three years.

The President, the Duke of Bedford, died in 1802 from a strangulated hernia which he had contracted while playing tennis on his own indoor court. He had been elected in 1799 and was active and influential in the Society's affairs to an extent that was, for Presidents, unprecedented. It was decided to commemorate him by means of a gold medal, to be awarded annually 'for agricultural improvement'. A subscription list was opened, to pay for the cost of the design and the die. Any money which remained after this purpose had been met was to be devoted to commissioning a bust or portrait of the Duke.

A special committee, known as the Bedfordean, was appointed to look after the matter and eventually chose a design submitted by a Miss Fanshawe, who was paid 21 guineas for her efforts. A Mr Milton, of Birmingham, was entrusted with the task of engraving the die, but the committee was far from satisfied with the first impressions taken from it, and set out their criticisms in a letter to Mr Milton.

'1st   We could have wished that the whole had been more relieved; in consequence of wh. the general effect wd. have been better, and the Bull which is the most distant object would admit to having been made more distinct.

2ndly We think the legs of the sheep too long by at least one third: this probably may be altered by cutting off so much of his Legs and raising the Ground on wh. he stands: by wh. means the distance will be precisely the same—at the same time the knee joint may be made made somewhat more distinct.

3rdly Some vestige of a penis should be given to the Bull. If you are unacquainted with the Forms of Cattle I beg leave to refer you to the Models or Engravings of Garrard.

13. Bust of the 1st Duke of Bedford, by Nollekens.

4thly The Border which you have sent and wh. I return is approved provided the
beards of the corn are omitted and each ear made somewhat longer in
proportion to their thickness. When the beards are left out there will be
sufficient space to admit of their increased length.'

Matters moved very slowly, however, and in June 1804 the Committee was
still requesting greater relief in the design. It was finally accepted in August when

14. The Bedfordean Medal.

Mr Milton was paid 100 guineas for the die and 20 to make a gold medal from it. The announcement had already been made in 1803 that the first medal would be awarded in 1804 for 'an Essay founded on practical experience on the nature and properties of manures', so the agreement with the engraver was timely. The first winner of the Bedfordean medal was Arthur Young. 200 copies of his essay were printed for distribution in pamphlet form, and subsequently in Vol. IX of the *Letters and Papers*. It was a substantial piece of work, and at the end of 100 pages, Young emphasised the need for farmers to understand the progress that was being made within the field of chemistry. 'Very few chemists have been farmers,' he observed, 'we can therefore do no more than combine the facts discovered by one set of men, with the results of the observations made by another set'. In this connection, it is interesting to notice the proposal made in 1805 by Sir John Cox Hippesley, that the Society should establish a Chemical Institution, a suggestion which was adopted at the Annual Meeting the same year. A subscription list was opened to provide the necessary funds and Dr Archer, who had proposed himself, was appointed, rather grandly, Professor to the Institution. He held the office for less than a year, death cutting short his Professorship.

The first Duke of Bedford was succeeded as President by the second Duke, who lasted only a relatively short time, resigning in favour of Benjamin Hobhouse, who had been a very active Vice-President, in 1805.

The arrangements for the Library were thoroughly overhauled in 1803. The special Committee listed 9 books which were known to be in the possession of members, but a further 23 which were missing, including 9 volumes of the *Annals of Agriculture*. It therefore recommended that in future 'no new publication whatever be taken from the Library within 12 calendar months from its introduction', and that 'no book whatever be taken without a Deposit being first made of the full value of such Book'. This tough line did not prove to be entirely workable and the rule about not borrowing new books was waived. The deposit system was, however, retained.

Two new Honorary Members were elected in 1804. One was Sir Humphrey Davy, and the other, more surprisingly, was Teyoninhok Arawen, otherwise known as Captain Norton, 'a Chief of the Mohawk Nation'. He was present at the Annual Meeting, and 'having been introduced to the sittings of the Anniversary by R. Barclay, Esq, rose and made a most appropriate address on the occasion'. The Society subsequently acquired a portrait of the Chief, in oils. It is now officially described as 'missing'. Barclay took a particular interest in the American Indians. In 1806 he sent the Society a pamphlet which contained an account of the 1795 Yearly Meeting of Friends in Pennsylvania, which had been mainly concerned with 'promoting the Improvement and gradual Civilization of the Indian Natives'. The Society was so impressed that it decided to send a gift of seeds, 'value not exceeding £5' to further this benevolent scheme.

Later in the same year, a resolution was adopted, recommending Phinias Bond, the British Consul-General in Philadelphia, as an Honorary Member, 'considering it would be to the credit and advantage of this Society to maintain a social and amicable Intercourse with foreign countries, particularly America'.

It is interesting to note, in the list of new premiums, an increasing tendency to include awards specifically for women. These covered a wide range of activities and achievements, but concentrated on work normally done by men. In 1800, for instance, Sarah Leonard, of West Lavington, received a 2 guinea award for reaping more than 5 acres of corn in a single day, but in the previous year there had been two even more remarkable feats. Hannah Mortimer ploughed 112 acres of land in the season—this won her 5 guineas—and Mary Masters reaped $8\frac{1}{4}$-acres of wheat in a day.

In 1805 a statement was published in Vol. X of the *Letters and Papers*, setting out in detail the amounts which had been paid out in premiums up to 1799, and adding the totals for each of the following years. This showed that £2,401 13s 0d altogether had been spent, made up as follows:

Up to 1799

'Agriculture, Planting, Cider-Making, etc.' (60 claimants)	£427 10 6
'Mechanicks and Arts' (19 claimants)	£109 5 6
'Gardening and Botany' (5 claimants)	£36 15 0
'Improving Cattle, Wool, etc.' (29 claimants)	£197 3 0
'Manufactures' (9 claimants)	£33 2 0
'Women Reaping and Hoeing' (9 claimants)	£18 18 0
'Essays on Agriculture, Planting, etc.' (9 claimants)	£76 13 0
'Friendly Societies' (3 claimants)	£31 10 0
'Prizes at Publick Trial of Ploughs' (19 claimants)	£60 13 6
'Labourers in Husbandry for bringing up large families without Parochial aid' (42 claimants)	£135 6 0
'Servants in Husbandry for long and faithful servitude' (117 claimants)	£317 12 0
'Servants for particularly meritorious servitude' (2 claimants)	£4 4 0
'Servants for excellence in Drilling' (2 claimants)	£5 5 0
	£1,453 17 6

Total for 1799	£122 17 0
1800	£172 4 6
1801	£174 6 0
1802	£130 4 0
1803	£209 2 0
1804	£138 12 0

It was noted that some members were of the opinion that too little was being spent on premiums. To them, the answer was made that the Society was obliged to live within its income. 'It is but fair to remark,' the paragraph went on, 'that such pressing proposals seem to be often made without considering the strength of a fund mostly composed of *guinea* subscriptions, and applied to many objects. It is the endeavour of the Society at its Annual Meetings to extend its encouragement as far as possible; but where gentlemen, after subscribing a guinea a year, are more intent on *the profit* of getting back twenty, than on diffusing *a variety* of useful knowledge, for the public good, they must be likely to suffer some disappointment, and perhaps the Society some unavoidable censure.'

The most ambitious project to qualify for a premium up to 1805 was for 'the most completely improved farm'. The specification shows the complexity of the problem with which the judges had to deal. The award was to be made 'to the person who, in the years 1802, 1803, or 1804, shall exhibit in one of the western counties the most improved and complete farm in respect to its implements of husbandry, appropriate buildings, plantations, fencing, draining, its various animal and vegetable productions, and management of cider; the value to be not less than 300L per annum, and the farm to be of a mixed nature, including breeding, grazing, and dairy stock; and the improvements to have been made by such occupier mostly within the last four or five years. A competition required, or the various merits of a single claimant to be of a positive nature—*Twenty Guineas*.'

There were four judges, headed by Billingsley, and, after a two day inspection, they decided that the farm of Mr White Parsons, of West Camel in Somerset, was worthy of the premium.

To the same volume, Billingsley contributed a well-informed assessment of what the Society had achieved since its foundation. 'It should seem,' he wrote, 'that the general opinion has been, that much public good has resulted from its exertions. If the truth of this position can be clearly shewn by facts: if the publication of its papers has produced happy effects in keeping alive a spirit of philosophic enquiry and political oeconomy; and the exertion of country gentlemen and farmers invigorated and extended; surely it must be acknowledged, that such an institution merits the support and approbation of every friend to national prosperity.

'Such a conclusion as the preceding seems to be supported by facts; and yet there are many, particularly among the classes of the less active gentlemen, mechanics of contracted views, and conceited farmers, who hold this and all similar establishments in utter contempt, ridicule their proceedings, and seize every occasion to manifest their disapprobation and dislike.

'Whether this ill-will proceed from an indisposition to contribute to its funds, or from a supposed conviction of its inutility, it may be difficult to determine; and perhaps the best answer which can be given to such hesitating and querulous characters, is that which a member of the Society gave to a farmer, who sarcastically remarked, that "He had been thinking whether the Bath Society had done harm or good?" "Have you," said our friend; "why, then, you may rest assured that it has done good." "Why?" rejoined the farmer. "Because it has led *you* to *think*, who seldom *thought* before".'

The root of the problem, as Billingsley saw it, was to persuade intelligent people to take an interest in agriculture. 'It seldom happens,' he believed, 'that a well-educated gentleman has a natural taste for agriculture. The ideas and habits contracted at the grammar schools, and the universities, are not very congenial to the laborious life of a farmer. Though a classical scholar cannot but be delighted with the description of rural employment and happiness to be found in ancient

15. Portrait of John Billingsley. From the Victoria Art Gallery, Bath.

authors, yet he seems rather disposed to see it through the medium of other people's practice than be a party himself.

'The appearance, the manner, the habits of the farmer, are by no means attractive; nor do I think, independent of other incentives, one in ten of those who have put their hands to the plough, would have so done, had they not been induced by the fashionable prevalence of agricultural conversation at the tables of the great, and the cordial and honourable respect paid to those who shine most in topics of this nature.

'A taste for experiment now animates the minds of most country gentlemen, and perhaps there are few things operate to encourage this taste more than the luxury of the times. A country gentleman, who some years since kept his carriage and his respectable table for 700L or 800L per annum, now finds himself in a straitened situation. He must either reduce his establishment, or have recourse to some means of increasing his income. Agriculture, therefore, naturally suggests itself as an honourable and effectual mode of increasing his funds. He takes upon him the task of a farmer, occupies in whole or in part his own estate, makes agriculture his study, and pursues it with unremitting ardour.

'His books inform him, that the greatest monarch, legislators, and statesmen, in former ages, considered agriculture and husbandry as objects peculiarly worthy their regard and encouragement.

'In his progress, however, he finds many difficulties occur; and such, in general, will ever be the case with this description of gentlemen; yet their disappointment and chagrin will commonly be attended with the effect of their lands being brought into a more valuable state for re-letting to labouring practical farmers, who, stimulated to industry, will produce more abundant crops for the common benefit of the country. But the grand source of increase in the county of Somerset, particularly, has been the abundant inclosure and cultivation of open, rough, and unprofitable lands, in which few counties, if any, may be found to have excelled this, during the period in which our Society has existed. Those improvements, if not exactly under the influence of the Society, have proceeded under its notice, and by persons who have taken a conspicuous part in its direction.'

Prominent among those who had 'taken a conspicuous part' in running the Society were the five members of the 'Committee of Superintendence', a creation of the 1790s. These gentlemen, who functioned in effect as a Council of Management, were elected annually. They were required to be people living near Bath, they had, we are entitled to assume, a fair amount of leisure, and they must on occasion have been a thorn in the side of the Secretary, especially a Secretary who disliked office work as much as Bartley did.

The Committee of Superintendence was particularly vigilant in matters of money. At the Annual Meeting of 1805, its members reported that 'they have had considerable trouble in the past year respecting omissions in the payment of Premiums decreed, and other Demands, even when the Sec. had sufficient monies in his Hands to have paid the same. To remedy this existing Inconvenience, in the month of July last your committee thought it right to draw on the Treasurer for £40 to effect the payments rather than suffer such Default to remain.

'And your com^ee also found a difficulty in approving the Sec's accounts by reason of some articles wh. we conceived to be unproperly brought into acct. against the Soc. We think it right to recommend that in future any compound quarterly charges for particular Expenses of Fire and Candle, cleaning Rooms, Porterage, etc. should cease to be inserted in these shapes and that all articles as they occur, and really cost, should be specifically charged in the accounts under the inspection of the Superintending Com.'

Chief among the extra charges which Bartley had put down were £52 10s 0d for attending Bedfordean committees, and £52 10s 0d for lighting, fires and cleaning from Lady Day, 1800 until Michaelmas, 1805. Both these items were disallowed, but Bartley, who had agreed to resign, was voted a gratuity 'not exceeding £50' for past services, if his accounts should ultimately prove satisfactory. On December 25, 1805, following the Annual Meeting, the Committee had written to him, requiring him to name a day on which he would present a full statement of his accounts. On February 22, 1806, he had still not done so, and a sharp letter followed. 'We entreat you,' it said, 'as you would regard your credit among your best friends, not to delay this business.' A postscript added, 'If you have two Bankers Account Books belonging to the Soc. wh. are missing from these rooms, you are requested to send them immediately to Mr Matthews'.

Eventually, the Committee got more or less what it wanted. The unfortunate Bartley—his offence almost certainly was muddle, not misappropriation of funds—furnished a list of subscriptions which he believed to be in arrears, some 'deeply in arrears', together with a complete list of subscriptions received but not entered—a total of about £50—which he was ready to pay in. In May 1806, the Committee, 'having taken into consideration the evasive conduct of Mr Bartley', sent him a strong letter about his continued failure to meet them and to settle his accounts. They threatened him with a public meeting, if he continued to disregard their efforts for a private settlement. In June they finally produced a total of £39 2s 7d which they believed to be due to them from Bartley, but there is no record, either in the Minutes or the correspondence, that this sum, or any part of it, was ever received.

It is very probable that the new President, Benjamin Hobhouse, was largely responsible for the decision to get rid of Bartley. A lawyer by training, and a member of an old Bristol family, he had a reputation for being both tough and just. At the time of his election as President, he was Member of Partliament for Grampound, in Cornwall, but in the following year he transferred his attentions to the Wiltshire constituency of Hindon, which he served until his retirement from Parliament in 1818. He was made a Baronet in 1812, and was also a Fellow of the Royal Society and a Fellow of the Society of Antiquaries. In the November following his election as President, Hobhouse presented 100 guineas to the Society. This, it was voted, should be invested, and the income used to provide a premium known as the 'Hobhousean', with a tablet recording the gift to be put up in the Society's rooms.

The new Secretary, Robert Ricards, or Rickards—the name is found spelt both ways—was a print seller and stationer. He was chosen by ballot from three applicants. At one time, the favourite candidate appears to have been a Mr Laurence, who had been recommended by Lord Somerville. Hobhouse wrote to him about the vacancy in December 1805, and at the same time Matthews sent a separate letter, outlining the Secretary's duties. The information supplied by

Matthews may well have frightened Laurence off—perhaps it was meant to—and we hear no more of him.

'The duties are very considerable in Number and Importance,' wrote Matthews. 'An extensive correspondence has been usually kept up, for the purpose of collecting and dispensing useful knowledge. The Number of Members living in various parts of the Nation are about 500. Their subscription accounts and the other accounts of the Soc. must be accurately kept. The Minutes and Transactions of the Meetings are to be regularly recorded, and the Business resulting from such meetings and Committees completed. The Library and Models are under the care of the Sec., who must spend 5 or 6 hours daily in the Rooms and not reside far from them. A volume of Memoirs has been usually published every two years, till the period of the last 6 yrs. during which the last Sec., through Indolence and Incapability has omitted that duty.

'The Salary is £100 p. ann. and no more. The new Chemical Department is to be kept detached from the office on wh. account no advance will be made. You will judge whether it could be adequate to your Time and Talents and the accommodation of a family, wh. you seem to be the Head of. I can scarcely suppose it can be. We reckon the Situat^n best adapted to a Gentleman of some small Fortune, and small wants. Many have applied, and two or three seem promising in several respects: but the situation will not be hastily filled up.'

A letter from Matthews to the Rev. J. Oldisworth, of Swansea, written a few weeks later, when the Bartley situation had proved even more difficult than at first supposed, confirms the impression that the Society intended to be ultra-cautious in the future. 'From the progress made in examining the erroneous accounts of the former Secretary', he wrote, 'such deficiencies are discovered as to confirm the opinion first formed that any Successor should be expected to give Security to the amt. of about £300.'

The new Secretary needed to be a paragon, in order to make up for the sins of his predecessor.

CHAPTER THREE

From an agricultural point of view, the really bad days may have begun with the end of the Napoleonic Wars, but industrial troubles had started considerably earlier. The Luddites were beginning their protest against unemployment and low wages among textile workers in 1811, and from then until 1818, when trade improved, agriculture and industry shared the national depression. Wheat prices went down from 126·6 shillings a quarter in 1812 to 65·7 shillings in 1815, when the first of the Corn Laws was passed, in an attempt to keep the price up. No wheat at all was to be imported until the price rose to 80/–, but by 1830 it had become clear that such a policy was futile, since the days of high war-time prices and high rents had gone for ever.

We now know that some rural areas came through these difficulties more easily than others. In general, the eastern counties, which were to a great extent dependent on cereal growing, did worst and the western stock-rearing counties best. What took place was not so much an agricultural as an arable depression. There were no riots in the West Country to equal those which took place in East Anglia in 1816.[1]

Yet—and such paradoxes are common enough in history—the foundations of great technical progress were being laid during the very years when the economic problems seemed worst. Stephenson's first locomotive ran in 1813, against a background of rural distress and Luddite arson and machine-breaking.

Experiment and uncertainty

The eleventh volume of *Letters and Papers* appeared in 1807. It was edited by Matthews, as Volume X had been. An article by Thomas Davis, on the Horningsham Female Friendly Society, is interesting both for the statistics it gives of maternal and child mortality in the area of the Longleat Estate—Davis was always quite as interested in rural people as in rural technology and rural economics—and for a note appended to it by Matthews, who refers to the article having unfortunately been omitted from the 1805 volume, owing to 'the general negligence wh. had prevailed during a late period, in the orderly keeping of the various MSS intended (as this was) for publication'.

An introduction, by Ricards, drew particular attention to the Society's new chemical laboratory, which has already received a brief mention. Dr Clement Archer's course of lectures, delivered in the Spring of 1806, 'was attended as well by many Ladies and Gentlemen who had a taste for the science, as by most of the Members who remained in town. These lectures were to have been occasionally resumed, particularly at those periods when the greatest number of the Members were likely to be on the spot. Unfortunately, death, which is continually depriving society of many useful members, bereaved this association of the zeal and talents of Dr Archer. The summer of 1806 put a period to the exercise of his private virtues and public exertions. Thus circumstanced, the practical part of the institution was left in the hands of the Committee above-mentioned, with the assistance of Mr BOYD, a very ingenious and intelligent chemist, who had been recommended by Dr Archer as his assistant operator in the laboratory, and whose real knowledge and acquaintance with the science is accompanied by that unassuming modesty generally attendant on true merit. The plan of the Committee has since been to procure soils, lime stones, and other substances from neighbouring (and in some instances remote) places, and to ascertain by analysis the nature and proportions of the component parts of such substances; thus enabling the intelligent agriculturist to apply his manure, and convert his limestones to use, agreeably to the indications pointed out by the result of any particular analysis.'

Mr Boyd himself supplied, as he had been asked to do, details of a do-it-at-home method of chemical analysis, but added a gentle and extremely tactful word of caution. 'With the view of promoting agricultural chemistry,' he reminded members, 'the Committee of Chemical Research have desired that a simple process should be detailed as an invitation to gentlemen to perform a few experiments for themselves; agreeably to which the following arrangement, by suspending rigid accuracy and facilitating operation, is recommended as introductory to complicated analyses. The knowledge of a few chemical combinations, practically obtained, may qualify the mind more clearly to understand chemical authors, and to comprehend the phenomena which take place in their various experiments.

'It must however be admitted that a perfect analysis absolutely requires the indefatigable application and extensive chemical skill which characterize the valuable labours of our eminent analyzers, to whose unwearied industry and splendid abilities chemistry is indebted for its important discoveries and scientific arrangements.'

In this volume, too, is Billingsley's *Essay on the Cultivation of Waste Lands,* which had won him the second Bedfordean Gold Medal. It contains a number of characteristically pungent remarks about his fellow farmers.

One should set about improvements with a complete disregard for any hostility or conservatism on the part of one's employees, an admirable sentiment which was easier to put into practice in the nineteenth century than it is today. 'A spirited farmer,' wrote Billingsley, 'should never be dismayed in the prosecution of experiments, by the reluctance or resistance of his workmen. Let him evince in his general conduct a prudent, discriminating, and steady judgment, together with a resolute and unvarying promptitude of execution, and he will always silence opposition, and endure obedience to his will. Whenever labourers are to be employed in any work or operation to which both they and their masters have been unaccustomed, one of the most confidential workmen should be set to work for a day or two, in order to explore the right way of proceeding, and to fix the contract price; and that he may do so with justice and impartiality, he should be promised a reward, if in the farther progress of the work, it appear that he has made a fair and correct report, and that the contract-men do not afterwards earn extravagant wages.'

The gentleman farmer's very gentlemanliness could be a serious problem to him. 'Buying and selling also constitute a very material part of a gentleman-farmer's care, and I scarcely ever knew a well-educated man competent to the task. The common practice of hacking and chaffering for a few shillings, and telling lies, either to lessen the value of the seller's article, or to advance the value of your own, is so disgusting to a liberal mind, that few gentlemen can reconcile themselves to the habitual practice of attending fairs or markets for any length of time. Here then your bailiff must be prime-minister; and if his ability and integrity stand the test, you are lucky indeed.'

In this vintage volume, there is an article by Matthews, explaining his proposal, accepted at the Annual Meeting in 1806, that the Society should have an experimental farm. '*A small experimental part of a farm*', he recalled, 'was attempted in the *infancy* of this Society, but on too contracted a scale, and under a system of management too defective, to be usefuly continued. Several such farms, of various extent, and under unfavourable circumstances, have indeed been attempted at different times by other societies, without success, and in more than one instance to the great disadvantage of the promoters; and hence seems to have arisen further discouragement at our board. But if *causes of failure*, from wrong principles of commencement and of management under speculative men, have been *obvious*, this should not seem to be a reason for our continuing to decline a laudable and useful attempt. Past errors in others should excite caution in us, and teach wisdom. There cannot surely be any sound physical reason why a farm belonging to more than one person should not be made *safe, productive*, and even in some degree *profitable*. Much must depend on the choice of the farm itself, much on the plan, but more on the *execution* of the plan.'

Matthews thought the Society itself should not run or own the farm. It should subsidise it, in exchange for an annual report and set of accounts. To finance the undertaking, 'Some One Member of the Society, of confessed ability and publick spirit, should be encouraged to declare himself ready to become a proprietor in a large proportion of the undertaking (say $\frac{1}{2}$ or $\frac{1}{3}$) and of course to become the chief director, as having an important stake; and that a very few others, equally well-disposed, and persons of his own choice or approval, shall be encouraged to join him, holding the remaining shares in such proportions as they shall agree on, the degrees of superintendance to be settled wholly among themselves. Such a plan of embarkation seems more desirable than under the sole management of one person, however capable; because, in case of illness or death, the publick advantage may be less likely to be lost.'

The proprietors 'shall be expected to have some one intelligent person as foreman or bailiff, resident on the farm, who may be authorised to attend occasionally, as *one* fixed day in a week, or otherwise, periodically, to explain on the farm to any gentlemen, Members of this Society, who may produce visiting tickets from the Committee of Superintendance, or from the Secretary: such tickets may also entitle the bearers to introduce with them any friend or friends whom they may be desirous to gratify. Non-subscribers not to have tickets alone. And the owners of the farm not to be subject to any irregular visits and interruptions.'

A Committee, which included Billingsley, was appointed to examine the possibilities. They recommended discovering 'a willing and suitable Farmer on a proper Farm'.

This excellent idea unfortunately came to nothing, but a number of more modest ideas met with greater success. Among them was a scheme for a proper exhibition centre in King's Mead Square, Bath. Here, the Society had available, in January 1807, an area 80 feet long and 32 feet wide, where they could develop reasonably good conditions for the Annual Show.[2] By April, they had put up sheds, and 'completed a good well of water', with a lease of 21 years to provide a reasonable measure of security. They 'formed all the rafters out of the Soc's own timber'. In February 1814, additional sheds were put up, at a cost of £98; foreign timber, it was found, would have raised the cost to £114. The Society needed the yard for such a small part of the year, however, that it seemed sensible to rent it out for the other months, and in December 1814, the buildings were leased to Richard Brooke for £40 a year, on the condition that the Society could have them for its own use whenever required and that Brooke carried out any necessary repairs at his own expense. The tenancy did not prove satisfactory, however, since by January 1818, Brooke was heavily in arrears with his rent. In May 1818, a Mr Selway was given the lease at £12 a year, which was probably a more realistic figure, but he likewise defaulted, having previously sub-let to a butcher. The next tenant was a Mr Hale, whose rent was further reduced in 1822 to 8 guineas.

The Society continued to pin its faith to premiums, as the best way of turning the energies of members in the most useful directions, and in January 1810, 'to give greater publicity to the views of the Society', 250 copies of the premium list were ordered to be printed and to be distributed as widely as possible. 'Several copies' were to be bound in stiff covers and placed, 'by permission of the landlords in the Public Room of the principal Inns of this City and elsewhere'.

The new premiums, not surprisingly, reflected the spirit and interests of the times. In November 1809, for instance, there was one for 'the best Treatise, either in Defence or Refutation of the system of the Rev Mr Malthus concerning population so that the objections which have excited so much alarm may be satisfactorily answered or his sentiments confuted'. The challenge was quickly taken up by William Matthews, who seems to have had plenty of time at his disposal, and his very long prize-winning essay was published in 1810, in Volume X of the *Letters and Papers*. The most interesting feature of the essay is not so much what Matthews has to say about Malthus but his opinion, strongly expressed here, that those whose business was the writing of history had not, for the most part, paid sufficient attention to the common people, and that the history which resulted was therefore unbalanced and inaccurate.

Malthus lived in Bath and there is a memorial tablet to him just inside the West door of the Abbey Church. But he does not appear to have shown any interest in the Society which took so much interest in him.

In 1799 winners of premiums had been given the choice of receiving their award either in cash or in the form of a piece of plate to the same value. Broadly speaking, the middle and upper classes took the plate and the poor went for the money. The system of presenting the premiums is described in a Minute of February, 1808. 'Agreeably to customary practice,' it reads, 'a Distribution of the Premiums awarded at the last annual meeting took place this day. Most of the successful claimants who chose Plate were present and received their Awards

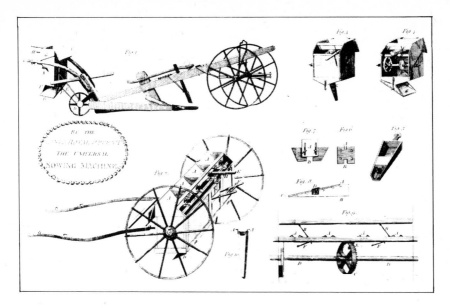

16. The Universal Sowing Machine. This shows the quality of illustration in the later volumes of the *Letters and Papers*.

at the hands of the honourable Baronet in the Chair. The Chief of the Premiums belonging to claimants of the inferior class[3] were deposited with the Sec. till called for.'

During the second decade of the century there were a number of changes affecting Hetling House, which, it will be remembered, belonged to Matthews. In September 1812, it appeared that the 'Soc's *Great Room* stood in need of *cleansing and beautifying* as far as relates to papering and painting, the Sec. was directed to appropriate the sum of £10', and in January 1813, a stove, produced and recommended by Dr Wilkinson, was placed 'in the upper part of the Great Room, to give that warmth thereto the want of wh. has been so frequently complained of'.

Matthews died in 1816, and in April of that year the Society made it known that it was 'desirous of paying this mark of Respect to his honoured Remains by attending his *Funeral*'. His executors requested permission to hold the sale of 'his valuable library of books' in the Society's rooms, and this was given. At the end of 1816, Thomas Davis, the son of Thomas Davis[4] of Longleat, bought Hetling House[5] from Matthews' executors and offered it to the Society on the same terms as before, agreeing also to pay the taxes on the house.

In June 1809, the Prince of Wales became Patron of the Society and it was announced that His Royal Highness would pay an annual subscription of 50 guineas. At the same time, the Second Warden of the Stanneries and the Surveyor of the Duchy of Cornwall were made ex-officio Vice-Presidents, in acknowledgement of their help in obtaining the Royal Patronage. The official letter of thanks to the Prince of Wales mentions 'the valuable and extended domains of the *Duchy of Cornwall* spread through the Western Counties and so closely approximating the Seat of the Society'. From that time until the present day, the Society has always maintained close links with the Duchy, and there is no doubt that these have been extremely valuable, both as a practical connection with one of the leading landowners in the South West of England and as a means of obtaining influential support for the Society's activities. It is worth remarking, in a similar context, that a number of Members of Parliament have been members of the Society and that one or two of them have provided an excellent two-way information service between Parliament and the Bath & West Society, often passing on reports of Committees to the Government, suggesting subjects for discussion and reporting on new legislation. At the beginning of last century, Sir John Cox Hippesley, M P, fulfilled this function admirably.

No time was lost in following up the advantages brought by the Prince of Wales' Presidency. In December 1809, an approach was made to the Lords Lieutenant of Somerset, Wiltshire, Gloucestershire, Hampshire and Devon, inviting them all to become Vice-Presidents. Thinking, no doubt, that what was good enough for the Prince of Wales was good enough for them, they all accepted.

There was some debate as to how the Prince of Wales' subscription might be most suitably and profitably used, and in December 1809 the decision was taken to devote it to the establishment of two premiums, the first to encourage the growing of turnips 'on the Forest of Dartmoor', and the second to 'a shearling Ram and 5 Ewes which appeared to be the most profitable, wool and carcase both taken into account'. It appeared later, however, that the Prince was a little disappointed about this, his own feeling having been that growing hemp, rather than turnips, would have been more suitable on Dartmoor. A letter to the Society early in 1810 from the Duchy of Cornwall agent, Benjamin Tucker, gives the total expenditure of the Navy on hemp during the recent wars, and emphasises that most of Britain's hemp was imported, particularly large quantities coming from Russia. The Prince of Wales believed that imports on this scale were both unnecessary and unwise and that, properly encouraged and instructed, British farmers could grow a great deal of the amount required.

The Royal displeasure would have been a serious matter, and the loss of the Royal patronage even more unwelcome. In 1810 it was consequently decided that the Prince's 50 guineas should be applied to stimulating the production of hemp, and that the premiums for turnips and sheep should be paid for from the Society's general funds. To improve relations further, all twelve volumes of the *Letters and Papers* were specially and handsomely bound and the President, Benjamin Hobhouse, presented them to His Royal Highness, as a token of the

Society's loyalty and esteem. The Royal set of *Letters and Papers* included the latest volume, which had just been published (1810). This, with great diplomacy, contained an article by Benjamin Tucker, 'On the Importance of Cultivating Hemp in the United Kingdom'. During the five years prior to 1809, Tucker pointed out, hemp imports from Russia had averaged £1,350,000 a year, of which £540,000 was for the Navy. This was apart from freight and insurance costs and 'profits of the contractors'.

'Having had the honour to lay all these statements before His Royal Highness,' said Tucker, 'he has been graciously pleased to command me to signify his request, that the Premium for the encouragement of the cultivation of Hemp may be so disposed of, as will best tend to invite competition *generally* throughout the united kingdom; in the anxious hope, that the patriotic exertions of the landed interest, under the fostering hand of Government, may soon enable the empire to dispense with all precarious foreign supplies of an article of such vast importance and magnitude of expence; and that, by liberal encouragement, the immediate and general cultivation of hemp upon the waste lands in the united kingdom which will admit thereof, may secure an ample and permanent supply of that indispensable article, upon which the prosperity of this country, if not its very existence, as a great independent nation, may eventually depend.'

The decision to encourage hemp-growing in England was something of a volte-face for the Society, since as recently as 1808 it had been pressing MPs to request the Government to find land abroad for growing hemp, seeing that the land in England was, in the Society's opinion, only sufficient for crops of grain and grass. By 'abroad' was meant 'some of the foreign settlements and perhaps in some of the lands in Ireland'. In any case, the Minute[6] stressed, hemp was 'an exhausting and precarious crop', a fact of which the Prince of Wales and Mr Tucker were perhaps not aware.

Volume XII also includes a report on a remarkable ploughing match, organised by the Society at Tetbury, in Gloucestershire. Three Beverstone ploughs were used, one pulled by two horses, one by a single horse and one by two oxen. The fastest work was done with the two horses and the best by the two oxen, but the efficient construction and light draught of the plough is shown by the fact that the single horse ploughed an acre in under five hours.

The plough was invented by Lewen Tugwell. The Society gave him a piece of plate, value 20 guineas, and 'in order to reward merit, excite emulation and remove prejudice, in every branch of the agricultural art, a silver cup, value four guineas, with a suitable inscription, be given to Thomas Pearce, the ploughman; and a coat of serviceable cloth made of British wool, with a set of the Society's new buttons, to John Perrott, who led the horse during the decision of Mr Tugwell's late ploughing match near Hunter's-Hall.'

This volume appeared late, because a year earlier, when it should have been ready, 'a sufficient number of communications had not been collected to form a Volume'. The main reason for this, the Society believed, was 'the number of public vehicles which furnish more speedy information, is much greater than it was when this Society first began to publish their Memoirs; and those vehicles are necessarily, on that account, resorted to by many *communicative* and *enquiring* agriculturists'. The Bath & West had shown the way, and thereafter there had been 'a great and growing increase of Agricultural Societies; composed, like this Parent Society, of Men of Noble Rank, of Gentlemen and Yeomanry, who, soaring above the dark regions of ignorance, revering no system *merely* for its antiquity, and despising the tyranny of deep-rooted prejudice, employ their talents in the cause of an enlightened system of agriculture; hoping to reap that ample reward which it is calculated to produce, in procuring necessaries and comforts for an increased population, and thereby ensuring the safety, independence, and prosperity of their country.'

A list of agricultural societies in the United Kingdom is given,[7] together with the places of meeting and the names of the Presidents and Secretaries, where these details could be ascertained. The total is 83, 49 being in England and Wales, 22 in Scotland, and 12 in Ireland.

At the Annual Meeting in 1810, it was agreed to publish future volumes of the *Letters and Papers* every three years, since a two-yearly interval did not appear to be realistic, and to make free copies available only to those who subscribed two guineas. Members who subscribed less than this were to pay half the selling price of the volume if, in fact, they decided to have it at all. This was not because the Society was in a serious financial position. On the contrary, the Committee of Superintendence had reported to the 1809 Annual Meeting on 'the improved state of its Funds, the Decrease in Arrears of Subscriptions, the increase of its Numbers and the growing state of its Prosperity'. This fortunate state of affairs allowed a generous gesture to be made to the Secretary during the following year, when it was agreed 'that the Sec. be permitted, at the close of every year's account to charge the sum of £20 to cover the deduction from his salary by the operation of the Income Tax and other items not necessary to be accounted for'. Income tax had been introduced to help finance the war with Napoleon: the Secretary belonged to the first generation of people who had to pay it.

The Society was increasingly worried about unemployment and rural poverty, and suggested in 1810 that some relief might be obtained by various means of improving the countryside. 'We mean,' said the Committee for Superintendence at the Annual Meeting, 'that the active moral benevolence in Country Districts may be particularly excited towards such Objects that as Idleness is the Source, at once, of Misery and Vice, some additional lands of employ in Field Labour may be contrived by the extra Levelling, dressing and particular Improvement of Cultivated Lands, Barren Wastes, Roads etc. etc. whereby wages may be earned, to the comfort of the employed and perhaps to the ultimate advantage of the Employer, thus lessening by provident means the Burden of the Parish rates and the growing Evils of Work Houses and in a word the Sum Total of human Infelicity.'

Among the other suggestions for easing rural miseries was one for subsidised milk for 'the Village Poor'. This was to be achieved 'by keeping Cows solely for the purpose of selling Milk at a reduced price to those poor families who have young children'. The proposal was approved at the Annual Meeting of 1806, but there is no report of the scheme being put into action.

Within the area served by the Bath & West Society, unemployment among textile workers was severe, and among the new 1810 premiums was one for giving 'useful labour' to unemployed workers in the wool manufacturing districts. One of the causes of misery among spinners and weavers was the cutting off of supplies of fine imported wool as a result of the Napoleonic war. In 1809, Dr Parry, in a note communicated to the Society, described the sheep-breeding experiments which he had by now been carrying on for eighteen years, and referred to 'the wretched state of the workmen employed in the manufacture of fine woollen cloth'[8] and took this to be 'an eloquent Proof of the indispensable Necessity of promoting the quickest possible supply of that valuable commodity'. He also pointed out that, in spite of all the research he himself had carried out, and the successful results he had obtained, the Board of Agriculture itself had given a Gold Medal 'to an excellent Manufacturer in a neighbouring County for a communication the object of wh. was expressly to prove that fine wool could *not* be maintained in this country by any Cross Breed of Sheep'. Stung by this failure of communication, the Society decided it should give wider publicity to the premium it had awarded Parry for his experiments.

Like all societies of whatever kind, the Bath & West had its blind spots, its prejudices, fads and its eccentric members. It is not clear, for instance, why Mr Lawrence of Birmingham was snubbed so forcefully in the matter of a picture of the Durham Ox. The brief facts, but not the background, are given in a Minute dated December 1808, when the Secretary reported, 'Read a Letter from Mr Lawrence of Birmingham soliciting permission to dedicate to the Society a print of the famous Durham Ox. The Society was directed to inform Mr Lawrence that the meeting did not chuse to comply with the proposal for fear of making it a precedent for future applications of the same nature.' In the following year, this decision was reversed, no reason being given. But, at this distance of time, the original request seems innocent and pleasant enough.

Another strange and unexplained incident took place in 1815. It concerned one of the Society's members, Robert Gourlay. It was reported at the Annual Meeting in these tantalisingly brief words. 'Mr Robert Gourlay having by advertisement, and placarding against the walls of Bath, made an attack, as gross and virulent as it was unfounded, upon the Society in general, and in particular upon what he termed the Executive of the Society, he was called upon to state whom he meant by that term, and explained that he did not mean any person in particular, but all the Members as a body. He was required to make an apology as public as his foul and unjustifiable aspersions had been, and declining to comply, he was expelled the Society.' But what Mr Gourlay plastered on the walls of Bath and what his 'foul and unjustifiable aspersions' were, we are left to guess.

One of the Society's interests for nearly forty years was what it described as 'the Neapolitan method' of killing animals. The idea was first put to Edmund Rack by William Hamilton, in a letter written from Naples and was still being canvassed in the 1820s. Hamilton supplied an explanatory sketch, 'a most exact representation of the knife with wh. the Neapolitan butchers kill oxen and sheep by separating the spinal marrow a little below the horns'. But, despite the offer of a premium and repeated assurances that the Neapolitan system was more efficient and more humane, British butchers continued to show an obstinate fondness for their traditional methods.

Two of the Society's most active and useful supporters, Sir John Cox Hippesley and John Billingsley, both died in 1811. Hippesley's services were acknowledged in a modest fashion, but Billingsley received a hero's treatment. 'The Society,' it was recorded,[9] 'willing to retain and perpetuate a grateful recollection of him and his services, unanimously voted that his Portrait should be placed in a conspicuous part of their great room; for which purpose leave was asked of his widow, Mrs Billingsley, to permit a picture in her possession to be copied. That lady politely acquiesced. Accordingly a very striking resemblance (painted by Mr Woodforde, and handsomely framed) is affixed in a suitable situation, with the following inscription over the picture:

'Non sibi sed toti genitum se credere mundo'

And underneath as follows:

JOHN BILLINGSLEY, Esq.

One of the original Founders, one of the greatest Ornaments, and for 32 Years a most able and active VICE-PRESIDENT, of this INSTITUTION, whose Ardour in acquiring Knowledge was only equalled by his Delight in imparting it; and whose Zeal in promoting Objects of public Utility was as conspicuous as his Judgment in discerning, and his Ability in carrying them into Effect.

The BATH and WEST of ENGLAND SOCIETY

In grateful Remembrance of his transcendent Merits, have caused this Tablet to be inscribed.

Mrs Billingsley made the Society a gift of £100, which she said was intended as a bequest from her husband and which would have arrived in the ordinary course of affairs, 'had he fulfilled his intention of making a new will'.

In 1812 the Bedfordean Gold Medal was presented to Sir Benjamin Hobhouse, as a token of respect and of thanks for his services to the Society. The previous year two silver Bedfordean Medals had been awarded for the first time. These were henceforth to be given as 'occasional Honorary Premiums', and the first recipients were Lewen Tugwell and Thomas Smith, 'for their great services to the Society'. Another Silver Medal was granted in 1812, to Dr Richardson, of Moy, Ireland, for his work in connection with the growing of Fiorin grass. The Society had come to the conclusion that turnips would never succeed on Dartmoor—Richardson refers to 'the elevated Bogs and chilling Fogs of that dreary Desart'—and felt that Fiorin grass might be a more satisfactory substitute.

In the same year, 1812, the Society's 'coat and buttons' was awarded for the first time. This was, in the first place,[10] for expertise in sheep-rearing, and was in addition to the premium of one guinea. The buttons[11] had been commissioned from C. F. Bullivant, of Birmingham, and from 1813 onwards the reference is always to 'the Society's buttons', although the buttons were in fact always presented complete with 'a stout coat'. The earliest description of one of these coats dates from 1819,[12] when a Mr Smith, of Smith and Crook, Devizes, attended with a coat for inspection. The Society gave him an order for 25, all the

17. The Bath & West Society's button, first awarded in 1812.

coats 'to be of the fullest and longest size . . . the charge not to exceed 2 guins. and including the *buttons* wh. the Sec. was directed to order from Birmingham immediately.'

The competitions for the sheep-shearing premiums were popular occasions and remarkable feats were achieved. In June 1815, it took Abraham Ford, 'who sheared his sheep in the best style of workmanship', 1 hour 33 minutes to shear his three sheep. The second premium went to Thomas Tyley, who took 1 hour 46½-minutes, and the third to Samuel Scrivens, who took 1 hour 39 minutes, 'but in an inferior degree of excellence'. On this occasion they also gave a gratuity of 7s 6d to James Weeks, 'for his attendance and laudable Attempt at Improvement', and 'a lad named Tyley, about 12 years of age, for exemplifying great dexterity in winding up the fleeces,' received a special gratuity of 5s.

In 1815–16 the Society's records contain many references to the distressed state of agriculture. At the Annual Meeting in 1815 a motion was carried calling attention to 'those Burthens wh. more immediately fall upon the Tenantry of the Kingdom; particularly the Tax called the Tenants Property Tax, the Taxes on Malt, Salt and Horses used in Husbandry'. An Extraordinary General Meeting, held in March 1816, noted with satisfaction that the war taxes on malt and on property were to be discontinued, but petitioned Parliament to introduce a measure of protection against foreign produce. 'It must be obvious to your Honorable House,' said the petition, 'that without the prospect of a fair and steady market for their Commodities, the agriculturists cannot employ their skill and capital to that advantage, which, by securing an improved produce, will render this country independent of foreign supplies, and thereby form the basis of its lasting wealth and prosperity.

'Your petitioners humbly represent that such advantages are not to be expected while the warehousing of foreign corn is sanctioned by law, together with the importation of cattle, wool, hides, tallow, butter and cheese, the produce of foreign countries, where the absence of tithes, poors' rates and other imposts enables the merchant to vend them at the British market at a price much below that at wh. our native cultivators can produce them.'

They asked for PROTECTION, and spelt out the word in capital letters.

Owing to ill-health, Sir Benjamin Hobhouse was not present at the 1816 Annual Meeting, and he resigned in the following year, his place being taken by Lord Lansdowne. In February 1818, the Secretary, too, gave notice of his intention to resign. He was, it appears, suffering from 'a paralytic affliction'. The Committee for Superintendence permitted him to continue in his office until Michaelmas, 'provided he will execute that office himself, without the ostensible assistance of a Deputy'. They expressed at the same time their 'unqualified sense of his (Ricard's) Integrity, Honor and Fidelity during a period of 12 years'.

This, however, is the last entry in the Minutes Book in Ricard's handwriting.

He was clearly too ill to continue, and in June Dr Wilkinson offered to act as Secretary until the Annual Meeting in December. Applications for the Secretaryship were invited, at an increased salary of £120, 'to include all charges for house and window-cleaning and coals wh. are in future to be defrayed by the Sec.' The Secretary was required to attend the office from 11 to 3, and to pay into the Treasurer's hands all sums received above £50.

The financial arrangements had already been considerably tightened a few months previously, when the Financial Committee reported[13] that in future accounts would be kept by the Secretary 'strictly conformably to the accounts of 1817 arranged and now produced by the Financial Com. as the best mode of presenting a clear and elucidated Statement of the Acc. of the Soc. and that the necessary books be provided accordingly'. The account books relating to this period have unfortunately not survived, but the volume of the Secretary's accounts for the period December 1796–March 1810 does not have separate pages for receipts and payments, which makes it difficult to find particular items and must have caused a great deal of confusion. The 'necessary books' referred to in the 1818 minute presumably made it possible to enter receipts and payments separately.

There were six candidates for the Secretaryship, to which Benjamin Leigh Lye was eventually appointed. He had been a Captain in the Army and seems to have had some small private income. The Annual Meeting which approved his appointment was notable for the large contingent from the nobility which turned out to support the new President. They included the Duke of Somerset, the Marquis of Bath, Earl Digby, Viscount Fitzharris, Lord Arundel and Lord Auckland. Lord Lansdowne celebrated his accession to office by presenting the Society with 100 guineas, $12\frac{1}{2}$ of which were spent on purchasing a cabinet for displaying chemical and geological specimens.

CHAPTER FOUR

*This chapter is concerned with the thirty years during which Benjamin Leigh
Lye was the Secretary of the Society, from 1819 till 1849. The national land-
marks of this turbulent period are easy enough to discern, at our present distance
from them. Set out as a series of dates, they would look something like this:*

1819 *The Peterloo massacre. The Six Acts, aimed at preventing revolution and
the growth of trade unions. The first crossing of the Atlantic by a
steamship, the 'Savannah'.*

1824 *The repeal of the Combination Acts, a token of Britain's emergence from
an atmosphere of slump and repression.*

1825 *Opening of the Stockton to Darlington railway.*

1830 *Agricultural labourers revolt against the conditions described by Cobbett
in his* Rural Rides. *Threshing machines destroyed, ricks burnt. Nine
hanged, 457 transported, many sent to prison.*

1832 *First Reform Act.*

1833 *First effective Factory Act.*

1834 *The New Poor Law established; relief to be given only in workhouses.
The Tolpuddle Martyrs transported for seven years for recruiting farm
labourers into a Union.*

1837 *Queen Victoria's accession.*

1839 *First Chartist Convention.*

1846 *Repeal of the Corn Laws.*

*But one could view the same 30 years as a pattern of trends and innovations,
rather than as a series of events. The availability of cheap tile-drains from the
1840s onwards transformed the methods of working heavy land, the steady
migration of families from the rural areas for factory work forced farmers to
think harder about the productivity of their labour force, and the need to provide
food for a rapidly increasing urban population made previous ideas about
restricting imports of foreign grain obsolete. Socially, the rural areas were not
pleasant places to live during the 'Twenties, 'Thirties and 'Forties. Technically
and economically, they were the scene of far-reaching changes, which some
farmers and landowners resented and resisted and others seized on and
exploited to the full, as the foundation of bigger and more prosperous enterprises. The Bath & West Society included, inevitably, both sorts among its
members.*

Into the doldrums

At the end of his first year the new Secretary was given a long memorandum[1] setting out his duties. The Committee of Superintendence proposed that, as the volume of business varied considerably, according to the season, his hours of attendance should be as follows:

November 1–March 1	11 to 3
March 1–July 1	12 to 2
July 1–November 1	12 to 1

The Committee had power at all times to grant him leave of absence, and its members expressed 'their entire approbation of the manner in wh. their business has been conduced by him (the Sec) and are persuaded that from his precision and zeal the Soc. will continue to derive, as they have derived, many important advantages. They are further persuaded that the duty is fully, or more than equal to the salary fixed under the present deductions and are of opinion that some arrangement may be made by wh. their officer may be a little relieved and the Society's interest as efficaciously supported.'

Translated, this meant that the Secretary, whom the Committee had learnt to trust, need not always be sitting in his office at the times prescribed, provided someone could be discovered who was willing to receive callers in his absence. 'A society consisting of Liberal Men,' the memorandum explained, 'should be supported on liberal principles, and your Com. firmly believe that business will be best conducted by those who, with the necessary ingredients of integrity and an acquaintance with the duties of the situation, possess also the spirit and feelings of Gentlemen. To encourage and retain such officers, some indulgence is due from a Society constituted like that of the B. and W.'

But not all the Society's members were, alas, gentlemen. One, a Mr Hunt, had to be expelled, after sending abusive replies to requests for payment of his arrears of subscriptions[2] and several others had behaved in a most irresponsible manner in the matter of prizes won by their employees. 'Several of the *great coats* so drawn as a reward and encouragement to their Servants have never yet been claimed by the Masters.' In one particularly disgraceful instance, the Committee reported, 'a Coat was drawn for the Servant of a member whose subscription was at the time 5 yrs. in arrear, he having never paid more than one guinea and that in another instance a Coat already delivered has actually been withheld in consequence of alleged ill conduct on the part of the servant'.

The new Secretary's first year was not a peaceful one. There had been a great deal of trouble in connection with Hetling House. The owner had refused to pay the taxes and arrears in respect of the rooms occupied by the Society. The Commissioners therefore issued a Distress warrant on the Society's property, and to avoid the execution of this, the Committee of Superintendence paid the £20 2s 6d demanded, 'intending to deduct the same from the next half-yearly payment of rent'.

Further trouble concerned the Society's collection of models and machines. The Committee of Mechanicks had inspected these, found many of them 'useless from natural decay and injury', and had them removed to the cattle yard. A catalogue of those remaining was, the Committee reported, in course of preparation, and new models had been purchased and put on show.

The Committee of Correspondence expressed concern with the lack of serious and worthwhile papers received from correspondents, such as were formerly available for printing in the Society's *Letters and Papers*. There is evidence in their 1819 report to the Annual Meeting that they felt that much of the contents of recent volumes had not been up to the required standard. To improve the situation, they believed that more good honorary corresponding members should be appointed.

The Finance Committee praised the 'meritorious exertions' of the Secretary, in having written 'near 300 letters' to persons in arrears, and by this means extracting £341 13s 0d from them. They thought it might help the Secretary in his future campaigns against members who failed to pay their subscriptions if he were allowed to style himself 'Sub-Treasurer'.

Once he had got the measure of the job, Leigh Lye made certain suggestions of his own for dealing with the Society's financial matters and accounts. He proposed giving £500 security, instead of the former £200, so that he could deal quickly with expenditure without having to wait for the Treasurer's draft at the Annual Meeting. He thought it was reasonable that he should be given such extra responsibility, since he was sure that the President and the Committee of Superintendence did 'not consider the Office of Sec. in this Institution as being held by a mere Clerk'.

His statement makes clear that the possibility of being allowed to conduct business of his own, as well as that of the Society, had been much in his mind when he sought the appointment, 'thinking it would be the means (if [1] conducted myself well) of obtaining other employments that would materially increase my income. I am sure,' he added, 'that there was a natural prejudice against me, from my having been so many years an officer of Dragoons.' He pointed out that 'under the present deductions' his salary for the year had in fact amounted to precisely £94 7s 2d and that nobody could be expected to live on that. The Society agreed to all his proposals, and raised his salary to £150.

This was in 1819, and a year later the Committee reported very favourably on the new arrangements.[3] They were, they said, well satisfied with 'the attentions of the Sec. to the multifarious functions of his office; and they do not find that any inconvenience or neglect of business has arisen from the remission of certain hours of attendance according to the indulgence granted him by the last A.M.' In March 1821, they agreed to allow him to transact the business of the Post Horse Duty for the City of Bath in the Society's rooms, and to take leave of absence periodically 'to enable him to execute the duties of Adjutant to the North Somerset Yeomanry'.

A more agreeable piece of business in 1819 was the arrival of the marble bust of Sir Benjamin Hobhouse, which had been commissioned from Chantrey. This was received with great approval. 'The superior execution of the *bust*', the Minute records,[4] 'excited the warmest approbation, it being thought not only to present a perfect resemblance of the universally respected Baronet but considered by connoisseurs as one of the most beautiful specimens of sculpture now extant in Europe'.

Sir John Cox Hippesley delivered himself of a 'just eulogy' on the bust. 'Accustomed as he had been to contemplate the best specimens of the Greek and Roman artists of sculpture,' he said, 'on the soil that gave them birth, as well as their rivals of the present age, he would venture to pronounce that there was no existing specimen of the art that could surpass that of Mr Chantrey in the present instance, whether it is regarded as a faithful representation of the Person and expression of the original, or itself as a chef d'oeuvre of sculpture in its execution'. The sculptor was considered to have deserved well of the Society and he was voted a silver Bedfordean medal.

Experimental work continued into the 1820s, although not on the same exuberant scale as ten or so years previously. Dr Wilkinson was as active as ever in and around his Chemical Laboratory and reported to Members, in 1819, that he could recommend, as aids to soil fertility, Droitwich salt, which was available in both Bristol and Bridgwater, and lime, 'through which gas was purified' from the recently established Bath Gas Works. A small consignment of rice seed was received from the Superintendant of the Botanical Gardens at Calcutta, who wished to discover if it could be cultivated in England, 'since it came from the mountainous districts of Nepal',[5] and a Silver Bedfordean Medal was presented to Rear-Admiral Bullen, 'the inventor of the improvement to the Ball Cock to check the waste of water'.

Ploughing matches continued—there were four during 1821—a model of a patent window-frame was demonstrated, 'which affords a facility of cleaning windows without risk or Danger',[6] and a ham cured with 'Pyroligneous or Wood Acid', was 'cooked on the Annual Dinner and placed on the table opposite the President'. A model of Fingal's Cave was given to the Society by a Mr Duncan,[7] the Rev James Plumtre sent a volume of songs called *The Experienced Butcher*,[8] and Captain Parry, RN, presented a stuffed musk ox from Melville Island,

18. Bust of Sir Benjamin Hobhouse, by Chantrey, 1819.

recommending that it should be placed in a glass case in order to preserve it.[9]

Premiums of five guineas each, with two guineas to be shared among the pupils, were awarded to schoolmistresses at Malmesbury and Cockroad, 'for instructing young females to knit'.[10] Most relevant to the development of agriculture and the improvement of rural life in the South-West was a resolution, adopted at the Annual Meeting in 1820, 'that a book be opened to receive subscriptions from such ladies as may be disposed to promote the improvement of small Dairy farms and the encouragement of meritorious female Dairy Servants'. Subscriptions received from this source were to be devoted exclusively 'to objects immediately connected with the management of dairies in the W. Counties, including the breeding of Poultry, Pigs, Bees etc.' Closely linked with this were the first Dairy Premiums, announced in December 1820, for the 'greatest relative produce of Dairy Goods for the market'. Dairies applying for these premiums had to have not less than four nor more than fifteen cows.

Excluding what was being done to improve the standard of dairy farming, most of what the Society was doing and discussing during the 1820s seems to have been distinctly lightweight and ornamental, compared with the pioneering and enthusiasm of twenty and thirty years previously. There were frequent motions deploring the state of agriculture, but a refusal to take any political action to improve matters. Parliament was still the unreformed Parliament, but the Society, as an organised body, seems to have held a remarkably high opinion of its wisdom and competence and to have been unwilling to criticise its actions or apply pressure to it in any way. Establishment-minded to a degree which the Society in the days of Rack and Matthews was certainly not, it was chary of supporting even the mildest petition to Parliament. This timidity was illustrated by the treatment given to William Hanning, a High Sheriff, in 1826. Hanning addressed the Annual Meeting of the Society and proposed a motion for petitioning Parliament on the Corn Laws. Various influential members, however, referred to 'the impropriety of agitating such a question in such a society'. Hanning then withdrew his motion and left the petition at the White Hart for anybody who might care to sign it. The 'noble President', Lord Lansdowne, welcomed this, and made known his conviction of 'the propriety of not entertaining any political matter in the discussions of the Society'.

The Society's finances inevitably reflected the depressed condition of agriculture. In December 1821, the Secretary begged leave, in view of the Society's shortage of money, to give up the extra salary which he had been granted in 1819. This offer was naturally well received and in their report at the end of the year the Finance Committee was very smug about Leigh Lye's 'zeal' and sacrifice. It recommended various economies and announced that it planned to use a legacy from Dr Fothergill, not for any constructive purpose, but to liquidate any existing debts. Playing down the financial saving that would result, the Committee urged that the practice of giving small sums to the unsuccessful competitors in ploughing competitions should be given up, as it occasioned as much irritation as satisfaction to those who had been hoping to win and was sometimes received 'with a degree of dissatisfaction and even of disrespect to which the Society should never subject itself in the distribution of any, even the smallest mark of distinction it may please to bestow.'

In the middle of its financial problems, the Society was forced to give urgent attention to the matter of Hetling House. In December 1823, the terms offered by

Davis for the lease were £75 for one year, £65 a year for three years and £60 (later reduced to £55) for five or seven years. The Society commented[11] that the premises were somewhat inconvenient, and such as they would not 'now perhaps engage for the first time'. On the other hand, 'they have answered all the purposes for wh. they have been required, for a series of years, and are moreover in a measure hallowed in the minds of the elder members by being associated with the recollection of the Rank and Talent, the Zeal and the Ingenuity exhibited in this spot during these years in wh. the Soc. was winning its way to the high Station it has since enjoyed amongst the Public Institutions of the Kingdom'. Conservatism and tradition prevailed, the Society made up its mind to stay where it was, and the lease was renewed for a further five years.

The Secretary, having agreed to give up a substantial part of his salary while the depression lasted, was compelled to look around for ways of supplementing his income. In February 1824, he was given permission to conduct business as Agent of an insurance company in the Society's rooms. This permission was confirmed in 1827.

19. Hetling House, the Society's first headquarters. Photograph, possibly by Fox-Talbot, c. 1850.

The Annual Meeting of 1826 considered a report from the Committee of Superintendence on the decline in the volume of the Society's business. The main reason for this, they believed, was the proliferation of other societies during the previous decade, as a result of which 'the business of this Society has not diminished in importance, but certainly reduced in quantity'. It was therefore possible, they thought, to deal with all matters requiring discussion at the General Meetings in November and December which paved the way for the Annual Meeting, instead of having these committees spread over the whole year.

There had been, the Committee said, an unusually large and interesting lot of essays and communications sent to them during the past two or three years, including a number for the premium on the planting of oaks. With no volume of *Letters and Papers* immediately in view, owing to a shortage of funds, it was decided to print 500 copies of the essays by Thomas Davis and W. Rogers, Nurseryman to the King, of Southampton, 'and to have the sketches lithographed'.

A further volume, No. XV, was, however, published in 1829. The introduction to the volume notes that thirteen years had passed since the publication of Volume XIV. During this period, 'the Society had been deprived of many of its most distinguished Members'. Of these, it remarks, 'some were connected with the earliest efforts of the Society, and all, by an unwearied and enlightened attention to its interests, contributed, in a high degree to the celebrity, which, during nearly half a century, marked its important progress. The loss of so much energy and talent would naturally paralize, to a certain extent, the proceedings of the remaining body. It happens, however, fortunately, that the history and progress of Societies, are analogous to those of the world in which we live. At certain eras, men congregate and apply their united talents for the relief of their mutual wants, and the advancement of their respective conditions. They make progress; the existing generation separates, or passes away; new generations arise, and the vigour of a perpetual youth animates the course of melioration improvement. In lesser Societies, though the objects of pursuit may vary with the change of time and circumstances, the hope of perpetuity can only be disappointed by the neglect or despondence of those who successively form their aggregate.'

The temper of the 1820s, the author believed, was different from that of twenty years earlier. 'The present,' he wrote, 'is not an age of theory and speculation. Mankind has been recalled by the extension of education, to a perception of the true principles of science, and the regular induction of facts. Experimental knowledge is in more esteem than at many former periods; and greater caution is used in the adoption of supposed truths, and questionable conclusions. It is evident, that, among the Members of the Bath Society, there are many practical and observing men, who, perhaps, from an unnecessary diffidence in this respect, withhold from the public inspection, the results even of judicious trials, and the suggestions of continued experience.'

Members' interests had changed, too. 'Since the foundation of this Society,' the article went on, 'very rapid and extraordinary advances have been made in many of the subjects which were, then, particularly offered to the consideration of its Members, and the acknowledged superiority of Great Britain to many foreign countries, appears at first sight, to connect less present importance with inquiries affecting its Arts, Manufactures, and Commerce, than at those earlier periods. With Agriculture, in some of its departments, the case is, however, widely different. Notwithstanding the manifest improvements which have long been taking place, this country has in some respects, scarcely yet attained the excellence which is discoverable in many foreign lands, whose cultivation has, appositely, been compared to that of a garden. This perceptible difference has very properly stimulated the energies of individuals and societies, and subjects of rural economy, which may be truly considered the source of all national prosperity, have attracted a more exclusive attention than in former years. Hence it has happened, that the proportion of those who contribute to the one, or to the other subjects, has materially altered: and that while the earlier Volumes of this Society are much devoted to communications respecting the Arts, Manufactures and Commerce, the later ones have been chiefly occupied by inquiries affecting the various branches of Agricultural employment. The communications on these heads have frequently been verbal, or have appeared as brief notices, connected with the exhibitions of live Stock, or of Implements, at General and Annual Meetings, the value of which have been duly appreciated, but which have been found incapable of adoption into a volume, as distinct Treatises or Essays. The materials for such a publication have, therefore, perhaps in the same proportion diminished, as the real practical importance of the Society has increased.'

In the publication field, the main problem appeared to be competition. 'Within these last twenty years,' it was pointed out, 'the number of weekly and of monthly periodicals, devoted particularly to these subjects of inquiry, has universally increased; and while, in some instances, pecuniary remuneration has attracted the competition of able writers; in all, the inducement arising from immediate insertion and rapid circulation has withdrawn papers from a work, of which the extension is more limited, and the publication deferred to a more remote period. In reference to the latter circumstance, the Society cannot but rejoice at the increased diffusion of knowledge, and, under such a calculation, can entertain little regret at its own diminished means of public estimation.'

It is curious, under these conditions, that the Society should have been so determined to keep its field as narrow as possible and, turning its back on the Arts, Industry, Commerce and Agriculture basis which was explicit in its constitution, to insist that it was an Agricultural Society above all things. This peculiar stubbornness is illustrated by the refusal to agree to Sir Benjamin

Hobhouse's wish that the premium associated with his name should be awarded for something connected with the Arts. The Committee of Superintendence noted[12] his request, first made in 1827 and repeated in 1828, but demurred, since 'insensibly and as it were by common consent its (the Society's) attention and its premiums were almost exclusively devoted to objects connected with Agriculture'.

Conservatism was probably the main cause of this decision, but linked with it was certainly the sad fact that the Society was so short of money that it felt compelled to channel whatever funds might become available into meeting the cost of existing premiums. There was, for example, the very worthwhile premium for the 'best managed and most improved farm'. This was awarded in 1828 to Mr Cook, of Down Ampney, near Cricklade, but at the Annual Meeting that year it was suggested that this particular premium should be removed from the Premium Book, unless some other means of financing it could be found. The possibility was mentioned of saving it by deleting a number of smaller premiums, especially those for bringing up a large family without the help of parish relief. Poor and hardworking parents who managed to do this should, it was felt, be rewarded directly by their parish, not by the Society.

The general gloom was relieved by the presentation by Mr Yarworth, of Trostry Lodge, Abergavenny, of 'a very handsome oil painting of a *singularly fine bull*, which might form an appropriate and pleasing decoration for the *Great Room*. The picture will have additional interest in the eyes of some worthy and respectable members when they learn that the animal it represents reached its enormous size without any other food than grass, hay and turnips. The Committee propose to affix a statement of his Breed, age and weight as soon as they can receive the particulars from his worthy Donor.' The bull's name was Trojan, and his picture is still in the Society's safe keeping. He was a pure bred Hereford, nine years old, and weighed $28\frac{1}{4}$-score.

But, in general, the Society's Minutes during the late 1820s give a sombre picture, both of the agricultural scene and of its own finances. Between 1818 and 1828 the income from subscriptions declined from £427 to £268 7s 0d. The Finance Committee recorded its belief that the number of members was unlikely to increase and urged the utmost economy. In 1829 it was proposed to hold the Annual Meeting at Bath and Bristol in alternate years, in the hope that the move would increase interest in the Society. The motion was carried, but by such a small majority that it was decided to withdraw it. One welcome piece of help came from the Corporation of Bath, which decided, in September 1830, to subscribe 20 guineas annually to the Society. The intention was, no doubt, good, but the Corporation soon discovered that its financial position was little better than that of the Society itself and in 1836 the Mayor wrote to say that, 'the supply having failed for the present', he was now making a personal donation of £10 and intended to take out a guinea subscription himself. As a token of goodwill, the Corporation agreed, also in 1836, to make the banqueting room of the Guildhall available, free of charge, for the Society's Annual Meeting and Dinner. Reporting this, with gratitude, the Secretary announced the cost of the Dinner was 'not to exceed 10/6 including a bottle of wine and the payment of waiters'.

The new Member of Parliament for Bath, I. A. Roebuck, did not follow the Mayor's example. The Secretary had written to ask him for a donation or subscription—his predecessors had always contributed £5 a year—but he replied that he was unfortunately unable to afford to subscribe, but was always willing to give personal help to constituents, an answer which must have brought the Society no comfort or encouragement at all.

During this period, certainly the most discouraging in the history of the Society, the nobility was not, perhaps, quite the tower of strength that might have been expected. In 1834, the Duke of Bedford wrote asking to be permitted to resign from the Society, not having been able to attend meetings for the past thirty years. The Secretary replied, deploring this withdrawal of 'your support in this hour of need', and his Lordship agreed to allow his name to stay on the roll. From 1830 onwards a number of Vice-Presidents resigned, mostly pleading the pressure of other business. The Society was always reluctant to let them go, knowing that it would have considerable difficulty in finding suitable replacements.

In 1834 a member wrote to say that in future he would not be subscribing to the Society, 'as he thinks it is not conducted as it used to be'. The Secretary was instructed to reply, saying that it would have been more satisfactory if the member had stated 'in what respect the Society is not conducted as it used to be'. The falling-off in support could not be debated away, however. At the Annual Meeting in 1834 it was not found necessary to erect the platform in the Great Room that had formerly been used to accommodate non-members. The atmosphere at Hetling House could hardly have been inspiring. 'The premises are in an extremely dilapidated state', it was reported in 1836, and the Committee of Superintendence was given authority to seek other accommodation should this prove necessary. Interest among Members reached such a low point that in 1836 the President, the Marquis of Lansdowne, wrote to say that he would take the chair at the Annual Meeting only if the Secretary could assure him that 'there would be a respectable meeting'. There was, he felt, no point in his attending if other 'gentlemen connected with the Society' failed to do so. In the event, he had, conveniently, a bad cold and failed to turn up.

The Royal Patronage fortunately continued with the change of Sovereign. In 1837, the Society wrote to congratulate the Queen on her accession, and received the news that Her Majesty would make an annual donation of £50 to the Society's funds. £50, however, was a mere drop in the ocean, and at the Annual

Meeting held in 1837 the Secretary stated that 'he had a communication to make to the Meeting which had been thought had better come from himself. Allusion had been made to the *thinness* of the meeting[13] and he was sorry to say there was a consequence, *thinness of the exchequer*'. He proposed giving up £20 of his salary, 'but he trusted he should not be thought guilty of presumption if he expressed a hope that the influential members would use their endeavours to increase the funds by prevailing on their friends to become subscribers'. He went on to say that 'no sacrifice on his part, even to the giving up of the whole of his salary, could keep the Society on its legs, or preserve its character amongst the number of similar Societies which had sprung up all around us, unless great exertions were used by those members who felt an interest in the prosperity of the Institution'. The meeting expressed complete confidence in Leigh Lye and thought his public-spirited offer to take a cut in salary would be unnecessary, at least for that year. It then awarded £177 in premiums, which was about the average amount and, curiously enough, more than in some of the prosperous years.

By 1840, the situation had not improved. As usual, financial troubles were blamed on the failure of members to pay their subscriptions, but bills had to be paid, and to allow this to be done the Society sold £300 of its investments. At the Annual Meeting of 1841, 100 new members were reported, but it was announced that, owing to a disaster for which the Society could not in any way be held responsible, drastic financial measures were inevitable. The Secretary had accepted a 50 per cent reduction in salary, the premium list was to be thoroughly revised and expenditure on the Library was to be severely pruned.

Before Fate struck, the position had been improving and a balance of £1,718 9s 0d had been expected. Then, the Society's bankers, Hobhouse Phillott and Larden, proprietors of the old Bath Bank, in Milsom Street, had failed, owing them £411 2s 9d, and it was impossible to say what sum 'may be saved from the wreck'. In the event, the loss was not quite as serious as had been feared. It proved possible to pay a dividend of seven shillings in the pound, which left the Society only £267, or nearly one year's income, out of pocket, but still able to hold on to what was left of its investment in Government stock. Tugwell, Mackenzie and Clutterbuck were unanimously chosen as the new bankers, but the problem for the moment was to find any money to pay into the account. It is a sad fact that a gift of £500 from the President, Lord Lansdowne, who was wealthy enough not to notice such a trifling sum, would have removed all the Society's worries, but neither the President nor any other of the great landowners made the slightest gesture of help. The books were finally balanced in 1844, when receipts had risen to £322 16s 2d, but the unfortunate Secretary was still having to get by on half-pay.

Faced with the amount of rural distress with which Members were surrounded during the 1830s and 1840s, it would have been surprising if the Society had not devoted a good deal of attention to the miseries of the agricultural labourers and to possible ways of improving their lot. The Annual Meeting of 1830 was occupied with receiving and discussing reports on this or, as they were described, *Essays on Alleviating the State of the Poor*. One of the most interesting of these reports came from John Beak, a farmer who had adopted the system of giving each year a piece of land in the turnip field to his workmen, so that they could grow potatoes for themselves. The men's potato-ground was 'ploughed and manured the same as for the turnip crop, 20, 25, 30 perches according to size of family'. This correspondent explained what the arrangements were after ploughing had taken place. 'In order that there may be no delay in planting,' he wrote, 'I take the whole of the men together, and make them help one another plant, for which day's work I pay them the same as though they were at work for me, and I convey the produce home for them without charge. In addition to this, I give those who have six children each a cottage and garden also rent free'. He had discharged only one man during the past ten years, and never altered their wages, these being from nine to eleven shillings a week. During harvesting and haymaking, they were paid piece rates, which amounted to more than their normal weekly wages. Mr Beak was commended for his plan and awarded the Society's Silver Medal.

20. Mr Winlaw's machine for removing grass from corn. From the *Letters and Papers*.

At the 1832 Annual Meeting the 'Agent', a Mr Perry, of the Labourers' Friendly Society, gave a long address on the advantages of giving allotments to the labouring poor. This society had just come to the notice of the Society. Based in London, it had 'high and extensive patronage', and produced 'frequent and cheap publications'. Whilst not unsympathetic to Mr Perry, the Committee of Superintendance thought the objects of his Friendly Society 'were being adequately catered for by this Society'. Members consequently needed to take no action in the matter, but were asked to take note of the existence and aims of the Labourers' Friendly Society.

From the many items in the Minutes, and from the awards made for virtue and long service, it is not difficult to construct a kind of Identikit picture of the Society's ideal farm worker. Sturdy independence would have been the keynote and the perfect example of what was hoped for was William White, whose case was presented to the Annual Meeting in 1832. 'At an early period in his life he began to save money, and at the age of 20 he built a stone house for the occupation of his parents, from whom he has never received any rent. At the age of 25 he had saved money enough to build a second house for his own occupation. At $25\frac{1}{2}$ he married and has since brought up 9 children, who are all unmarried and live with William White. Neither he nor they have ever received relief or assistance from any Parish, not even during the years of scarcity. He now occupies about $4\frac{1}{2}$ acres of land upon which such of his children as do not find employment as day-labourers are employed, and the ground is well stocked and in good cultivation. William White is a sober, moral and religious man and has brought up his children in habits of honest industry.' There is, it will be noticed, no mention of Mrs White, who must also have possessed qualities above the average.

If the Society had felt able to do more than award premiums to encourage other labourers to be like William White, if it had not been so anxious to stand

---

THEATRE-ROYAL, BATH.

UNDER THE PATRONAGE OF THE

MOST NOBLE THE MARQUIS OF LANSDOWNE,
And the BATH and WEST of ENGLAND AGRICULTURAL SOCIETY.

This present TUESDAY Evening, December 8, 1840,

WILL BE PERFORMED MORTON'S ADMIRED COMEDY OF

Speed the Plough.

Bob Handy .... Mr. BENSON

Sir Abel Handy ... .Mr. JOHNSON	Henry...... ....Mr. H. HOLL
Sir Philip Blandford ....Mr. RANSON	Morrington.......................Mr. GLANVILLE
Farmer Ashfield .....·............Mr. WOULDS	Peter ....Mr. WILSON
Evergreen ....Mr. BETHWAY	Gerald ....Mr. WEBB
Postillion ,..........................Mr. KIMBER	Bob Handy's Servant ....Mr. LAWLER
Susan Ashfield.....................Miss LACY	Lady Handy...................Miss HIBBERD
Miss Blandford...................Miss MORVIN	Dame Ashfield..............Mrs. W. H. ANGEL

21. Theatre bill for *Speed the Plough*, performed at the Theatre Royal, Bath, 8 December 1840, during the Society's Annual Meeting.

well with the land-owning interests, it might have transformed itself into a powerful moral and political force during the dismal period of the 1830s and 1840s. As
it was, it contented itself for the most part with platitudinous sentiments and
accepted a markedly defeated and defeatist position. Its great misfortune during
these years was that its members included no powerful and colourful character of
the Billingsley, Hobhouse or Matthews type, no-one with the courage to persuade the Society to strike out boldly in unfashionable directions. Nothing illustrates this better than the fate of a series of resolutions concerning the administration of the Poor Law. These were proposed at the Annual Meeting in
1830 and seem, to anyone reading them today, both constructive and non-
revolutionary. But there was no strong and influential person to push them
through, and, after a long discussion, they were not adopted.

What, briefly, the Society proposed, but lacked the courage to advocate
publicly was as follows:

1. that the agricultural worker should receive a wage which would be adequate to
   allow him to obtain the necessities of life, 'and also, under strict economy, to
   provide for the wants of a family which may be dependent on him';
2. that, in addition to wages, he should have 'that small additional remuneration
   which would be necessary to provide, under a system of mutual assurance, either
   voluntary or compulsory, a portion of such earnings during the time of
   sickness';
3. that single men should be paid the same as married men;
4. that no relief, in the form of money, should be paid to men who were unemployed.
   Instead, there should be some provision of work at the expense of the parish, but at
   a lower rate, for these 'surplus hands remaining unabsorbed by the natural
   channels of employment';
5. that there should be allotments for agricultural labourers;
6. that the above proposals should be recommended to landowners and to
   Parliament.

These excellent and modest ideas were quietly buried, and all that the Society
felt able to offer as a substitute was a new premium to encourage landowners to
let land, in plots of not more than half an acre, to 'the labouring class'.[14]

It would be misleading, however, to suggest that, during what can now be seen
to have been the Society's worst years, the 1830s and 1840s, when it could easily
have disappeared altogether, nothing of any value was accomplished. Some of
the premiums, such as those awarded to children for straw-plaiting, may appear
trivial, but this was an age when any addition to the family income was welcome,
and, apart from the profitable skill involved, a prize of 20 shillings to a 13-year-
old was a small fortune. Many of the inventions placed on exhibition at Hetling
House were ingenious, and in their way important, but not of a type which would

get the farming community wildly excited and anxious to subscribe to the Society. One could mention in this connection Dr Wilkinson's apparatus for heating conservatories and his fire escape;[15] Mr Higman's apparatus for use in the event of shipwreck, 'a life-preserver which he had invented, consisting of basket-work in which were seven pieces of cork or old bottle corks';[16] a model of a suspension bridge, from Cadogan Williams;[17] Mr Buckland's scheme and model for cast-iron caissons, filled with rubble, to protect the coast against erosion.

As time went on, it became clear that, on many farms, further improvements were impossible unless something could be done about the drainage, and this became the Society's main campaign during the 1840s. In 1844, 1,000 copies of the details of a new £100 drainage premium, awarded by Lord Lansdowne, were printed and distributed at Wells, Shepton Mallet, Frome, Devizes, Warminster, Westbury, Chippenham, Calne, Melksham, Bradford-on-Avon, Trowbridge, 'etc.'. This produced very little response for the first year and Lord Lansdowne absented himself from the Annual Meeting of 1846, as a sign of his disappointment and disapproval. In 1847, however, the Society recommended that it should be awarded to the sole claimant, Mr Hulbert, of Ford Farm, Bradford-on-Avon.

The premium was originally for improving drainage by means of a subsoil plough. The Committee thought it might be better to have £50 for that and £50 for any other method, and that perhaps one premium of £30 and two of £10 would produce a larger entry than a single premium of £50, but his Lordship was not willing to consider any further subdivision, and the premium, as it eventually took shape, was for two awards of £50, the first to a farm of not less than 100 acres and not more than 20 miles from Bath, which improved its drainage by any system, and the second for a similarly located farm which had drained not less than 15 acres by means of a subsoil plough. The small amount of interest this produced was remarkable. Farmers in the Bath area did not seem to be drainage-minded.

One can see, from a study of the Minutes, that the Society was apprehensive about the effects of the recently founded Central Agricultural Association, regarding this as a threat to local societies in general and itself in particular. The establishment of this national society was reported at the Annual Meeting in 1835 and in the following year the Committee of Superintendence endorsed the Society's refusal to support the policies of the Association, referring darkly to its 'political character'. The Central Agricultural Association quickly failed, to be succeeded by The Farmers' Central Agricultural Society, which appears to have been an equally crusading (i.e. 'political') body, not unlike the present National Farmers' Union. It planned to make representations to the Prime Minister every week or every fortnight. In 1839 it called on all local societies to affiliate, and sixty did so within the year. The Bath & West, however, refused to do so, saying, in a distinctly prim and holier-than-thou manner, that it was 'a society for scientific and practical improvement in agriculture', and that, if there had to be a national society, it preferred the a-political Royal Agricultural Society, which had just been founded (1838).

The Leigh Lye era came to an end in 1849, and no Secretary to the Society was so sorely tried for so long. During February and March, 1845, he was too ill to attend meetings, and, although he dragged on for another four years, the task of making arrangements for transferring the Society's headquarters from Hetling House was more than he cared to contemplate. At the Annual Meeting of 1848, it was recommended that the lease should be terminated, since not only was the rent high but the building was in a very poor state of repair. The Bath Literary and Scientific Institution was approached, with a view to sharing premises, but eventually the decision was taken to re-settle the Society in rooms in Church Street, near the Abbey, which would, it was felt, be sufficient for all purposes except the Annual Meeting.[18] A subscription for the Secretary, who had held the post for thirty-one years, raised £185. The Society resolved to give him two salvers, worth £40 each, and the rest 'in a purse'.

In July 1849, the Committee of Superintendence met to consider Leigh Lye's resignation. They regretted this very much and said that 'looking moreover at the diminished patronage and support extended to the Society and the lamentable falling off of the attendance of members it becomes a matter of serious consideration whether the Society may not be said to have accomplished the object for which it was instituted and whether under the combined circumstances just mentioned the time has not arrived for dissolving the Institution than that it should be a lingering death of comparative uselessness.'

In August the Committee summoned a special meeting of Vice-Presidents to consider the matter, and this meeting, somewhat surprisingly, declared its confidence in the Society and its prospects and refused to agree to the quick death recommended by the Committee of Superintendence. A new Secretary was sought, and the Committee recommended that 'this distinguished position shall in future be an honorary one'.

Henry St John Maule, of Bath, undertook the duties on these terms.

CHAPTER FIVE

This chapter deals with the 1850s, when the Society came to life again. It ac-
complished this against a background of great social changes and advancing in-
dustrial prosperity in Britain as a whole.

The Corn Laws were repealed in the year Benjamin Leigh Lye resigned as
Secretary. Two years later, in 1851, the Great Exhibition gave an enormous
stimulus to inventiveness in every field and did a great deal for the morale of the
British people at the same time, although the enormous progress made by the
Americans, much in evidence at the Exhibition, was not to everyone's taste. The
publication of Darwin's Origin of Species *in 1859 did as much to crack the crust*
of intellectual and religious conservatism as to propagate scientific ideas. The
American Civil War, in 1861, and the cutting off of supplies of cotton which
resulted from it, reminded Britain in particular that even the most important in-
dustrial country in the world could not exist in isolation.

But in agriculture changes were taking place which rivalled those in industry.
It is, perhaps, wisest to mention people first. In 1851, 18 per cent of the occupied
population of England earned a living from agriculture; in 1871 the figure had
gone down to 11 per cent. It is true, of course, that the total population increased
considerably during the same period but, even so, farmers were employing con-
siderably fewer people at the end of the 1860s than at the end of the 1840s, and, at
the same time, producing a great deal more food. The efficiency of the industry, in
other words, was increasing all the time. Previously uncultivated land was being
brought into use, machinery was taking the place of men, more fertiliser was being
used, more meat was being produced. And, at the same time, food prices fell
dramatically by 25 per cent between 1846 and 1849. In the mid '60s, the index of
food prices was still 12 per cent below the level of the mid '40s.

The great revival of the Bath & West Society took place against this
background.

The Acland Revival

After the Annual Meeting of 1849, the Society began to dispose of many of its possessions. The portraits were something of a problem. Mr Davis took that of his grandfather, but it proved difficult to trace any relatives of John Billingsley. Eventually, a Mrs Seymour, living at Yarmouth, in the Isle of Wight, was discovered to be Billingsley's only surviving child, and she said she would be happy to receive her father's portrait. Sir Edward Parry wrote to give the Society authority to do what they liked about the Musk Ox. The Commercial Institution offered to take charge, for the time being, of anything for which a home could not be found and this suggestion was gratefully adopted.

One very sensible decision was taken at this time. In 1849 the President, Lord Portman,[1] informed the Society that he was always engaged at the Smithfield Show and at the meeting of the Royal Agricultural Society at the date of the Bath & West's Annual Meeting. Would it not be possible, he asked, to change the date of the Society's Meeting, to avoid this happening in future, since he was extremely anxious to attend? The Society agreed, after having refused to make the change almost since its foundation, and arranged that, in future, it would hold its Annual Meeting during the week preceding the Smithfield Show.

Further administrative decisions were taken at the same time. The Committee of Superintendence was henceforward to consist of 24 members, at least a third of whom were to be members paying the minimum subscription of half-a-guinea. All the sub-committees were to be appointed from the members of this Committee. A big effort began to secure new members, especially from industry and commerce, sources which, it was felt, had been unreasonably neglected in the past.

Within two years, a very marked improvement had taken place in the Society's affairs. In 1851 a merger with the Devon County Agricultural Society was arranged and in 1853 the fact that the Society had taken on a new lease of life was emphasised by the publication of the first of a new series of *Journals*.

The organisation and membership of the Society at the beginning of its revivified period is worth giving in some detail. The officers were:

'*Patron*
The Most Noble the Marquis of Lansdowne

Vice-Patrons
The Duke of Wellington
The Duke of Somerset
The Duke of Bedford

The Earl of Mount Edgecumbe
The Earl Fortescue
The Earl Digby
The Earl Fitzhardinge
The Lord Portman

President (for the year ending June 1852)
The Lord Portman, Lord-Lieutenant of the County of Somerset

Vice-Presidents
There were 57.[2] Of these 25 were titled and 11 Members of Parliament. There were also the Lord Warden of the Stanneries, The Surveyor-General of the Duchy of Cornwall and the Receiver-General of the Duchy of Cornwall.

The Council
24 members from the Eastern Division.
24 members from the Western Division.'

There was a Publication Committee, with five members, and a Finance Committee, with three. A network of 24 Correspondents was set up, 12 in the Eastern District and 12 in the Western. These had promised 'to use their influence in extending a knowledge of the Society and increasing its income. They have been authorized to receive subscriptions. It is much to be desired that a local correspondent should be found in or near every market-town, as the punctual collection of subscriptions is one of the most urgent necessities, and, unhappily, one of the greatest difficulties of Agricultural Societies.'

Two treasurers were appointed, one for the Eastern District and one from the Western. These gentlemen, members were informed, 'undertake the trouble gratuitously'. In 1852 the number of subscribers had reached 420, their contributions totalling £550 16s 6d. To this had to be added a further £150 from donations and life subscriptions.

The 1853 volume includes a report of the results of one of the Society's most momentous decisions, to take the Annual Meeting away from Bath and to hold it each year in a different town within the Society's area, combined with a Show of machinery and livestock. The first of these peripatetic Shows was at Taunton, which contributed £200 towards the expenses.

The programme of this momentous Taunton meeting has survived. It is printed on two foolscap sheets:

PROGRAMME

MONDAY, JUNE 7th.   Last day of receiving Implements, &c., to be exhibited in the Implement Yard, and arranged by the Stewards for the Judges Inspection.

TUESDAY, JUNE 8th.   Stock received into the Show Yard, until Seven o'Clock p.m.

At Six p.m., or as soon after as the Judges of the Implements shall have completed the trials, Members of Council and Governors of the Society may enter the Implement Yard, Tickets 2s. 6d. each.

WEDNESDAY, JUNE 9th.   The Implement Yard open to the Public from Eight a.m. to Six p.m., Tickets 2s. 6d. each.

The Judges to inspect the Live Stock and award the Prizes; and at 3 p.m., or as soon as the Judges shall have delivered in their awards, the Public to be admitted into the Cattle Yard at the entrance in Barrack-street only, Tickets 2s. 6d. each.

>   N.B. *Notice will be posted over the entrance when the Judges have completed their awards.*

THURSDAY, JUNE 10th.   The Cattle and Implement Yards open to the Public from Seven a.m. to Six p.m., Tickets 1s. each; and from Six p.m. until Eight p.m., Tickets 6d. each.

The Public Dinner to take place in a Tent in the Vivary Park, at half-past Three or Four punctual. Tickets 2s. 6d. each.

In the Evening, a Ball at Meetens's London Hotel

TICKETS:—GENTLEMEN, 10s. 6d. and LADIES 7s. 6d. each, including Wine and Refreshments. Tickets to be obtained at the Hotel, and at Mr. SUTTON's, Bookseller.

FRIDAY, JUNE 11th.   The Cattle and Implement Yard open to the Public from Nine a.m. to Four p.m., Tickets 6d. each.

The Live Stock may be cleared out of the Yard before Nine a.m., and if not then removed, must remain until Four p.m.; the whole of the Implements will be on view, and such Livestock as may remain.

The Cattle will be shewn in the Barrack Yard, and the Implements in a Field adjoining.

The admission to the Implement Field (on the morning of Wednesday) will be from the entrance in Mount-street; after the Cattle Yard is open, from the entrance in Barrack-street through the Cattle Yard.

PRESIDENT:

THE LORD PORTMAN, *Lord Lieutenant of the County of Somerset*

STEWARDS OF DEPARTMENTS:

Cattle   J. WEBB KING, Chilton Polden;
                                    C. GORDON, Junr.;

	W. PORTER, Hembury Fort.
Implements	HENRY PARAMORE, North Petherton;
	HENRY BLANDFORD, Orchard Portman;
	SAMUEL PITMAN, Rumwell, near Taunton.
Director of the Show	THOMAS HUSSEY.
Local Secretary	J. LEVERSEDGE, Bath Place, Taunton.

By Order of the Council,
H. ST. JOHN MAULE,
May 11th, 1852. *Secretary.*

Office of the Secretary in Taunton, during the Exhibition, at MR. DAVIS'S, Carver, &c., High-Street

By the Regulations of the Society:— ·

All Persons admitted into the Show Yards, or other Places in the temporary occupation of the Society during the Meeting, shall be subject to the Rules, Orders, and Regulations of the Council.

GENERAL CONDITIONS

STOCK

No charge will be made for Hay, Vetches, or Straw for any Stock by whomsoever Exhibited; from the time of entry into the Yard, to Four o'Clock p.m., on Friday the 11th.

IMPLEMENTS

Exhibitors are required to mark the *lowest selling Price* upon every Implement, &c., exhibited in such a manner that it may be conspicuous to the Public.

Horses will be provided at the expence of the Society, for the trial of Implements.

There will be an Exhibition of Poultry, Rabbits, &c., and Certificates of entry may be obtained of the Secretary, Bath; or of Mr. LEVERSEDGE, the Local Secretary, at Taunton. The entry closes on the 22nd of May, 1852.

The Vale of Taunton Deane Horticultural and Floricultural Society, will have a Show on Wednesday and Thursday, June 9th and 10th, in the Vivary Park.

Should duty not prevent, it is expected that the Royal Marine Band from Plymouth, by the liberal permission of the Commanding Officer, will be allowed to attend.

No Money will be taken at the Entrances, and the admission will be by Ticket ONLY, to be obtained at the Shops of Messrs. MAY's, SMALL, SUTTON, BARNICOTT, BRAGG, COURT, ABRAHAM, DYER, and HISCOCK, Booksellers, Taunton.

Subscriptions and Donations may be paid to MR. R. BADCOCK, *the Bank, Taunton.*

In adopting the policy of going from town to town in this way, the Bath & West Society was doing no more than follow the example of the Highland Society, which had been doing the same thing since the 1820s, and the Royal Agricultural Society, which began its county shows at Oxford in 1839.[3] In 1850 a prominent member of the Bath & West Society, William Miles, wrote a penetrating assessment[4] of the new type of show and of the ways in which it could be integrated into the educational work of a local society. In 1848 the Royal Agricultural Society had brought its annual show to Exeter and those who attended this, wrote Miles, 'must, I think, have felt overwhelmed by the sight of so many implements which had never found their way into the west; and must have wished for further opportunity of examining their merits. . . . It was a great matter for us in the West, to have Garrett and Hornsby, and Busby, and Crosskill brought from Suffolk, Lincolnshire and Yorkshire to our very doors; but when shall we see them again? We must take some steps to supply our local wants for ourselves, for the enormous expense of freight must hinder the ordinary farmer from going to them for what he wants.'

Miles therefore proposed an annual peripatetic show in the West of England. He thought this would have three advantages:

'First, it would give some encouragement to the Ironmongers in our Market towns to take pains to find out the most acknowledged improvements and to buy the articles on speculation.

'Secondly, competition would drive out of the market certain trade articles which find a sale in most places from mere habit for want of being confronted with better things, and a public trial would test the working of novelties merely got up for sale, or articles fraudulently pretending to be what they are not.

'But the greatest benefit of all would be the improvement of our local Wheelwrights, Smiths and Carpenters. There are among them many ingenious men, and good workmen, but they want extended knowledge, and a higher standard of excellence in point of Workmanship. I was informed at Exeter, I do not know how correctly, that a very clever workman of Thorverton lost the prize for his waggon only because he was not aware of the improved method of making wheels, by which an immense saving of draught has been effected without any loss of strength.'

The Bath & West Society decided to adopt Mr Miles' plan. The prize money totalled £484—cattle £148, sheep £99, pigs £30, horses £40 and implements £167, 'which last item includes a prize which is a new and striking feature in agricultural exhibitions, peculiar to this Society, being the prize of 20£ for the most economical collection of implements suited to tenants occupying arable land not exceeding 100 acres'.

The judges were given these instructions:

'That they be requested generally to bear in mind the objects of this society, namely—to benefit the agriculture of the West of England, and therefore that those animals and implements should be encouraged which are suited to the soil, climate and other peculiarities of the district.'

'With regard to cattle and sheep and pigs, not to take into consideration the present value to the butcher of the animals exhibited, but to decide according to their relative merits for the purpose of breeding; therefore, that particular attention should be given to those points which indicate a tendency to produce offspring with healthy constitutions, with due regard to symmetry, size, and such other points as afford the best prospect of profit.'

'With regard to horses, to consider especially the qualifications for farmers' work in a hilly country, whether as agricultural horses or as hackneys.'

'With regard to implements, to bear in mind that in the West of England farms are generally of small extent; to give especial attention to small implements for the cultivation of green crops and for the preparation of food for stock, and generally to give the preference to implements of simple construction, light weight (with due regard to strength), handy for use in a country in which stony ground is very common.'

There were 126 exhibitors, 77 of stock and 49 of implements. Of the stock exhibitors, 40 came from West Somerset and 22 from North Devon. A similar pattern was shown by the implement makers, 13 of the 49 coming from West Somerset and 13 from North Devon.[5]

'An important feature of the Show,' members were reminded, 'was the Catalogue, containing a full account of all the stock and implements exhibited, extending to nearly seventy pages, and sold for sixpence. Every farmer would do well to purchase this document, not only for use in the yard, but to preserve it for future reference.'[6]

The report on the display of implements at the Taunton Show contains some interesting general remarks.[7] 'Presuming then,' the authors[8] begin, 'that all intelligent agriculturists are satisfied that the shows of the Royal Agricultural Society have in great measure caused the rapid progress in the improvement of implements, we proceed to state the additional advantages likely to result from a local show if it is properly supported.'

These advantages were:

1. As 'a bazaar or market for the sale of goods of a peculiar kind, not easily accessible in any other way without considerable expense and trouble. Unlike the commodities of domestic life, they are not readily to be found for inspection in the ordinary course of retail trade, nor can they be ordered from common country artisans, because their production requires considerable mechanical knowledge, excellent workmanship, and an acquaintance with practical farming in its most minute details. The purposes for which they are required are so various, that no manufacturer, unless he is a man of extraordinary skill and capital, can hope to supply the best article in every department; and improvements in construction are so rapidly made, that retail shopkeepers are afraid to speculate largely in articles which may in a short time be superseded. The result is that the most celebrated makers of agricultural implements are to be found at great distances from each other, in some of the most highly farmed districts.'

Manufacturers from all over Britain brought their goods to Taunton, because 'a show-yard well attended by discerning purchasers is the best and cheapest of all advertisements'.

2. Farmers bought because they could see the goods demonstrated.
3. Local ironmongers were willing to stock 'those implements which gained prizes or obtained a ready sale'.
4. ... 'the stimulus given to the local smiths and wheelwrights; perhaps this operation of the show was not altogether palatable to the exhibitors who came from a distance, and who think themselves obliged to charge prices sufficient to cover the cost of invention. Several of our local artisans, having not in vain visited the Crystal Palace or the former shows of the Royal Agricultural Society, produced articles very nearly akin to the most celebrated prize implements, at very low prices.'
5. ... 'the benefit which local makers may derive from the light thrown upon the defective construction of their implements by the trials, and from the advice given by the judges as to the best mode of overcoming the defects so made apparent.'

The new-style Bath & West followed the example of the Royal Agricultural Society in its *Journal* too. Since its first publication in 1839, the Royal Society's *Journal* had aimed at two things:

'1. To record the actual practice of good farmers, by collecting facts from all parts of the empire.

2. To push forward the inquiries of science, with a view to ascertain the principles involved in every-day practice, and by fresh applications of those principles to simplify and cheapen the processes previously in use.'[9]

For its own *Journal*, the Bath & West Society announced that it was seeking contributions under four headings:

1. 'Local climate and soil'
2. 'Reports of practical farming'
3. 'Scientific principles so far as applicable to practice'
4. 'Unsettled points on which information is needed'

The papers printed in the 1853 volume followed this general pattern:

1. Lord Portman, 'On the General Principles of Agriculture'.
2. William Miles, MP, 'On Mangold Wurzels'.
3. Philip Pusey, 'On the Progress of Agricultural Knowledge during the last Eight Years' (reprinted from *Journal* of R.A.S.).
4. Nicholas Whitley, 'On the Climate of the British Islands' (Prize Essay) (reprinted from *Journal* of R.A.S.).
5. Thomas Dyke Acland, jun. and Samuel Pitman. 'Report on the Exhibition of Implements at Taunton'.
6. T. D. Acland, jun. 'Report on the Exhibition of Live Stock'.
7. Gabriel Poole, 'Remarks on a Trial of a Reaping Machine'.
8. Ph. Pusey, 'On Nitrate of Soda as a Top-dressing of wheat'.

A section called *The Farmer's Note Book*, contained notes on such subjects as 'Comparative Merits of Teams of Oxen and Horses', 'The Potato Disease', and 'On the Effects of Coprolites and Guano on the Growth of Turnips'.

The balance sheet for the year ending November 1, 1852, shows how the Society was receiving and spending its money at this time.

'A.

Balance from the United Societies of Bath and Devon	£163	13	11
Donations	120	5	0
Life-Compositions of Members	60	0	0
Annual Subscriptions	519	8	6
1½-Year's Dividends from Devon Fund to July 1852	26	3	4
1 Year's Dividends from Bath Fund to July 1852	20	7	10
Subscriptions from Taunton	210	0	0
Produce of Taunton Show	482	2	11
	1,602	1	6

B.

Liabilities of Society previous to its extension	46	1	7
Subscription to Bath Institution	1	1	0
Printing, Advertising and Expenses (sic) of Meetings of Provisional Committee and Council from Dec. 1850 to July 1852	170	9	0
Secretary $\frac{1}{2}$ Year's Salary to July 1852	26	5	0[10]
Show-yard Erection and Expenses	541	3	10
Judges	49	12	11
Society's Prizes	430	0	0
Director and Local Secretary	21	0	0
Expenses of Committees of Inquiry on Disputed Prizes	9	15	0
Balance in hands of Treasurer to 2nd November 1852	279	5	3
Cash in hands of Secretary 3rd November 1852	27	7	11
	1,602	1	6'

Although the remodelled and re-invigorated Society could not feel really secure until the mid-1850s, when it had become clear that the new policies were working successfully, a great deal of useful planning and experimenting was done from the first months of the new Secretary's appointment. Attendance at the Annual Meeting in 1850 was little better than it had been for some years past, only the President, nine of the Vice-Presidents and twenty-five Members taking the trouble to turn up. 1851 produced double that number, many of whom went on to the Annual Dinner at the Greyhound Hotel, at 3s 6d a head. But the decision to take the Annual Meeting and Show round the region transformed the scale and complexity of the Society's operations. The organisation of the Show demanded managerial ability of a calibre not previously required.[11] There had to be negotiations with local authorities, transport of machinery and stock had to be arranged, large numbers of people and animals had to be fed, temporary buildings erected and cleared away speedily.[12] A completely new set of skills had to be learnt. The Council Minutes for 1852, for example, contain interesting entries describing an approach to the railways, to discover if they would convey stock and implements for the Show either free or at reduced rates,[13] and an enquiry to a number of the principal towns in the region—Plymouth, Devonport, Totnes, Newton Abbot and Torquay—asking if they would be willing to follow the example of Taunton and make a contribution to the expenses of the Annual Meeting, if it were to be held in their locality. The amount of money handled each year had to be reckoned in thousands, instead of hundreds. Correspondence became much more extensive, a serious matter in the days when every letter had to be written by hand.

In five years, the Society was hauled, more or less awake and willingly, out of the eighteenth century and into a steam-powered, scientific world.

The negotiations with the railways showed how much sound commercial sense had been learnt in such a remarkably short time. 'The great and constant liberality of the Bristol and Exeter and South Devon Railway Companies,' the Secretary reported, 'has been invaluable to the Society. The Directors of the Great Western Railway Company have kindly acceded once more to the earnest request for concessions to the Exhibitions of privileges similar to those granted to Exhibitions at the Royal Agricultural Society's Meetings. The Council cannot help expressing a hope that they may think it right still to continue their support to an object of so much utility for the sake of the stimulus it gives to general business, and they will consent to transmit once a year, on reduced terms, the sample of goods, of which the bulk is transmitted through the year at full prices. The Council are in possession of important statistics, by the manufacturers, showing that a considerable extent of traffic along the principal lines of railway is traceable to the annual exhibition in the West of England, established on a scale known to no other Society. It is far from improbable that this recently extended traffic would in a great measure cease if impediments were placed in the way of the annual attendance of the implement-makers for the purpose of receiving orders.'

This revival and the re-introduction of a sense of relevance and urgency into the Society's affairs was due mainly to the efforts of two men, the new Secretary and Thomas Dyke Acland. Their temperaments could hardly have been more different, Maule being exceptionally meticulous and orderly and Acland brimful of ideas, but without the patience and administrative ability to see them through to the end. They made an excellent pair.

Acland was born in 1809, seven months before his life-long friend. Mr Gladstone. The Aclands, with 15,000 acres, were among the largest landowners in Devon. The future 11th Baronet went to Harrow and then to Christ Church, Oxford, where he took a double first in classics and was then elected a Fellow of All Souls. From that moment on, his life can be not unfairly summarised in a single penetrating sentence once written ⌐bout his father—'Without a profession, he seemed to have qualified himself for several, and he laboured for himself and his friends' friends as if he were called to the work by the most urgent professional duty or interest.'[14]

He did most of the things which were expected of a man of his position in the county—Parliament, the Volunteers, the Church—but he tended to do them in a somewhat unconventional way. His career as a Member of Parliament illustrates this very well. When he was twenty-eight he was elected as a Conservative for West Somerset. At the General Election of 1841 he preferred to be returned as a Liberal Conservative and in 1846 he committed at least temporary political suicide by voting again for the repeal of the Corn Laws, a thoroughly unpopular thing to do in a rural constituency. Having lost his seat in Parliament as a result,

22. Sir Thomas Dyke Acland, as a young man. Pastel portrait, in the National Portrait Gallery.

he said farewell to politics for the time being and went to King's College, London, to learn chemistry. This was not a matter of self-indulgence. He wanted to be in a position 'to show the farmers a better method of improving their position than was to be found in any reliance on Protection'.[15] The phrase 'show the farmers' had real meaning in Acland's case. He wanted to teach, not preach. In 1855 the Society published[16] a 48-page article by him, called 'Elementary introduction to the Chemistry of Practical Farming', and explained that it was 'a lecture delivered in the parish of Broad Clyst, with additional matter'. Broad Clyst, in Devon, was his home territory and no doubt the lecture had been given to the family tenants and other interested people in the locality.

Acland was out of Parliament for eighteen years and during this period he took a very active part in the affairs of the Society—he edited the *Journal* for seven years—and in educational matters, particularly those affecting the West of England. He also raised and trained five corps of Mounted Rifle Volunteers in Devon, acting as Colonel of the 3rd Devonshire Volunteer Rifles from 1860 until 1881, and Major of the 1st Devonshire Yeomanry Cavalry from 1872 until 1881. He wrote a manual of rules for the Volunteers.[17] It was a characteristic document. The system of fines contained in it illustrates his passion for precision and for getting things right.

'For loading contrary to orders, or shooting out of turn 2s 6d
For discharging the rifle accidentally 5s 0d
For pointing the same, loaded or unloaded, at any person without
    orders 10s 0d'

There is an interesting paradox here. On the one hand, he disliked sloppiness of all kinds. It has been truly said of him that 'in things agricultural it was the rule of thumb that was his enemy. . . . He would urge farmers to get rid of slovenly work and slovenly opinions, to think of the reasons for what they did.'[18] But on the other hand, 'his manner of working was curiously disorderly. All sorts of important matters were delayed and put off in favour of the reigning topic of thought.'[19] Why was the author of the *Rules of the First Devon Mounted Rifle Volunteers* apparently unable to write and follow a book of rules for himself?

There seem to have been two main reasons. The first, not infrequently found among men of power and wealth, especially inherited power and wealth, was the comforting certainty that there were always other and lesser people at hand to tie in the ends which one left loose oneself. And the second reason was that he was interested in so many things, and that there was never sufficient time for all of them. He knew so much and he threw off ideas, often big and important ideas, almost as easily and naturally as he breathed and ate.

'His talk was apt to be a little overwhelming,' wrote someone who worked closely with him for many years, 'owing to his complete absorption in any subject uppermost in his mind, and to the velocity with which he poured forth his thoughts; this sometimes lessened the effectiveness of what he had to impart.'[20] The flow of his ideas was unending, and the Bath & West Society benefitted greatly from them. He had the presence, the personality and the energy to match the ideas. 'His very appearance always commanded attention. A certain rugged picturesqueness of face and feature, with a light in the eye that on occasion was full of humour, was set off by a strikingly fine physique. Clad in a rough home-spun suit, with a satchel of agricultural literature for distribution slung low down at his side, he would, with some brief spells of rest in the Secretary's office, stride about the Bath & West show yard all day with an uprightness of body and a power and elasticity of gait which many a man twenty years younger—for he could then be seen when he was close upon eighty—might envy.'[21] It is the portrait of an evangelist.

He had a thorough and first-hand knowledge of farming in the West Country, more particularly in Devon and Somerset. Soon after losing his seat in Parliament, he decided to compete for a prize offered by the Royal Agricultural Society on the farming of Somerset. To make sure of his facts, and to get his proportion and emphasis right, he spent a great deal of time travelling all over the country, visiting farmers, landlords and agents and gaining a first-hand knowledge of agricultural problems which stood him in good stead in later years. Two things impressed him above all others, and he stressed them in his report, the need for farmers to have much greater security—'No man can farm well unless he can look with confidence beyond next Michaelmas'—and the very unsatisfactory conditions under which a great many, if not most, of the farm workers were compelled to live. He won the prize, and his essay was published in the R.A.S. *Journal* in 1850. He had, incidentally, been one of the founders of the Royal Agricultural Society and became a member of its Council as early as 1838.

Under Acland's editorship, the *Journal* of the Bath & West Society became once again something it had long ceased to be, a volume of practical use to farmers and, with more than 300 pages for each issue, it provided excellent value for money. In his introduction to the second volume, published in 1854, Acland was able to point out with pride that nearly all the contributions came from the Society's own members. There were essays 'On the most Economical and Profitable Method of Growing and Consuming Root Crops', by W. C. Spooner, of Southampton, J. T. Davy, of Roseash, Devon, and J. Webb King, of Chilton Polden, Somerset. George Arnold, of Dolton, near Crediton, received a prize for his Plan of Double Cottages, with specifications. 'The separation of the sexes is provided for in the bed-rooms', Mr Arnold pointed out. 'The ascent to the bedrooms is from the living-room, in order to give more warmth up-stairs; but there is a fire-place in the principal bed-room, an important point both in case of sickness and for general ventilation.'[22]

But Acland felt it was important, even if articles by members had to be excluded, that articles originally published by the Royal Agricultural Society should be given a second readership in the Bath and West Society's *Journal*. 'In a district such as ours,' he believed, 'no greater service can be done to the rising generation of farmers than to lead them to that great storehouse of information, the *Journal* of the Royal Agricultural Society, by reprinting papers which they might otherwise never have the opportunity of reading.'[23]

Great efforts were made to increase the circulation of the *Journal* and to make it a propaganda weapon for the Society. 'The Journal Committee,' it was reported, 'have great pleasure in acknowledging the fact that above seventy

PLAN OF DOUBLE COTTAGE
FOR FARM LABOURERS.

FRONT ELEVATION.

END ELEVATION.

SCALE OF FEET TO ELEVATIONS.

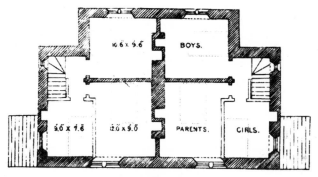

10.6 × 9.6	BOYS.
9.6 × 7.6 12.0 × 9.0	PARENTS. GIRLS.

BEDROOM PLAN.

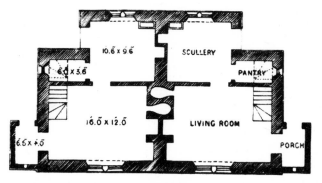

10.6 × 9.6	SCULLERY
6.0 × 5.6	PANTRY
16.6 × 12.6	LIVING ROOM
6.6 × 4.6	PORCH

GROUND PLAN.

SCALE OF FEET TO PLANS.

H.W. HICKES, *ARCHITECT.*

23. Plan of Double Cottage for Farm Labourers. In *Journal*, 1856.

copies of the *Journal* have been ordered for gratuitous distribution among their tenants by owners and managers of property within the West of England'. Anyone who was willing to buy at least six copies for this purpose got them at a reduced price. In the Report of the Council there is an account of the fact of one thousand copies printed of the first issue of the *Journal*.[24]

'Members' copies	428
Copies sent to contributors, to newspapers, to local correspondents, etc.	141
Sold to the public	21
Supplied to new members	45
	635'[25]

But the *Journal* absorbed about a third of the Society's annual income, which was, in any case, much lower than it should have been, because members were regrettably lax about paying their subscriptions. The office in Bath was far away, and memories were short. Acland wrote feelingly about this and explained how the Society proposed to deal with the problem. The answer was something that would nowadays be called administrative decentralisation.

But, five years later, he had to admit that, despite a gratifying growth in the number of members, the Society's income was not increasing to the extent that was necessary. The Committee wagged a firm finger at the members. 'After giving a copy of the Journal (delivered gratis),' it said, 'and a free Ticket of admission to the yard to Subscribers of Ten Shillings, little, if anything, accrues to its funds from their subscriptions, the Committee also state their belief, that many respectable yeomen subscribe only ten shillings, under the impression that that amount is the usual rate.'[26]

Men like Acland—and there were a few more of the same calibre in other agricultural societies—got innovations accepted by sheer determination and strength of personality. They found local agricultural societies a useful vehicle for their energies. 'Their aim,' if we have to be reminded, 'was to pump new life into the countryside, and in this sense to civilise it. Housing, education, industry, employment, religion, all came within their field. The local agricultural society, with its mandate to interest itself in every branch of human activity, was a perfect field of operations for these many-sided men.'[27]

The *Journal*, under Acland's direction and inspiration, reflected this determination to report on and assess new ideas wherever they were to be found. But the emphasis was on ideas. 'The Journal Committee,' members were told, 'is most anxious to record the observations of practical men, and to gather useful hints from all quarters',[28] but observations and hints were not everything. 'We think,' the article went on, 'that one chief end to be aimed at by writing is to stimulate inquiry into the grounds and reasons of things; to explain and diffuse a knowledge of first principles, which, if they are true, will lead to further improvements of practice.'

So science was mingled with technology, the why with the what. A good example of the method can be seen in the report of the Exhibition of Implements at the Bath Meeting in 1854. 'Year by year,' it begins, 'have the machines produced been more and more suitable to the peculiar wants and necessities consequent on local circumstances, and these are nowhere more strikingly apparent than in the strongly-marked features of the western counties. Year by year have our local mechanics had their energies tried and their talents stimulated by being brought into competition with our most celebrated agricultural engineers.'[29]

This 46-page report devotes 16 pages to examining in great detail the performance of six steam-engines, with the aim of distinguishing between real and nominal power. By showing farmers why one engine was superior to another, it was hoped to make clear to them what did and did not constitute value for money. But without the dynamometer, invented by W. Froude, 'Civil Engineer, Dartington, Member of the Council', the experiments would have been impossible. To have a professional engineer on the Council was in itself a sign of the times.

Acland took a keen and discriminating interest in contemporary art and artists. He was one of the early admirers of Millais, bought his work and figures in two of his sketches. Another of his friends was Ruskin, and in 1871 Acland and William Francis Cowper (afterwards Baron Mount-Temple), were the original trustees of Ruskin's Guild of St George. It is not surprising, therefore, that he should have worked hard to establish an Arts Committee of the Society—he was its Chairman—and an Arts Section at the Annual Show. In 1860 the Show was at Barnstaple and for the first time an Arts and Manufactures section was added to the usual exhibits of livestock, machinery and equipment. The aim was to give visitors to the Show an opportunity—quite probably for the first time in their lives—to see a collection of good paintings and sculpture, together with examples of contemporary industrial design. The Department of Industrial Art at South Kensington (the forerunner of the Victoria & Albert Museum) provided an exhibition of various types of 'industrial art', mostly ceramics at Barnstaple, and other material was lent by manufacturers. The fine arts section included some drawings by Turner, 'some of whose relations were actually living in Barnstaple as artisans and came to look and wonder.'[30] A special weather-tight and lockable building, 100 feet long and 40 feet wide, with a wooden floor, had to be made available. It was in three sections, with the Fine Arts and the South Kensington exhibits sandwiched between two groups of 'artistic manufactures'.

The Society took the experiment very seriously, believing that rural people would benefit from a chance to see something better than the shops and stalls of the local market town normally offered them, and that it was as important to try to raise the level of public taste as to improve the standards of horses, pigs and agricultural implements. Two principles were to be followed in selecting the exhibits of 'industrial art'. The first was 'that ornament should be consistent with the construction required for real use, not the construction made to fit to a pattern chosen for the sake of ornament, just as a horse's shoe should be made to fit the foot, not the foot be fitted to a ready-made shoe.' The second guide-line, equally admirable and necessary, was that 'the natural qualities of the material should be considered in framing the design, whether for construction or ornament, the opposite error being very common, namely, to borrow patterns invented by the workers in stone, wood or iron, and attempt to reproduce the pattern in some material for which it was not intended. Each material has its own natural qualities, and therefore its own style.'

On the first occasion, in 1860, there were 'good collections of cutlery and other hardware, of philosophical instruments, decorative ironwork, bronzes, fire-arms, carriages, furniture, etc.', together with numerous examples of glass, pottery and lamps, 'branches of manufacture to which England, during the last few years, has paid great attention, and in which it has indisputably made much progress.' The exhibitors included Price's Patent Candle Company, with a display showing candle-manufacture and the ingredients used.[31]

Older and choicer pieces of pottery came from South Kensington. Visitors were told that 'with the assistance of the catalogue, and under the guidance of Mr Worsnop, the intelligent and obliging curator', they had a splendid opportunity of 'making acquaintance with the Majolica, Fayence, Palissy, and most of the other wares which have at various times been produced in Europe, as well as to see genuine specimens of Oriental manufacture'.

The Fine Arts section at Barnstaple included, as well as the drawings by Turner, 'spirited sketches, or carefully elaborated studies from Nature', by living artists from the West of England, who were assured that 'by sending their works for exhibition at an agricultural show, they did great good and gave a great deal of pleasure; by placing fine art in the midst of decorative art and manufacturing art, as helping to improve the taste of the manufacturer and the purchaser, they vindicated their own true position. They have a message to deliver to the agriculturalist and to the manufacturer.'

At Dorchester in the following year, the exhibits included cut glass, by Pellatt of London, 'made expressly for the use of his Imperial Majesty the Sultan, at his palace on the Bosphorus', Holman Hunt's painting, 'The Strayed Sheep', lent by Mr Charles Maud, of Bathampton House, and several pictures by Frith, from the collection of Mr W. C. Lacey, of Wareham. 'The Strayed Sheep' received particular praise from the catalogue. Among its other merits, 'to each animal is given an individuality which must have been thoroughly appreciated by the sheep-farmers of Dorset.'

24   William Holman Hunt: *The Strayed Sheep* (1852), exhibited at the Dorchester Show, 1861. Now in the Tate Gallery.

Sculpture was not represented at Dorchester, although it had been at Barnstaple. 'The heavy and bulky package of works in marble and plaster renders the carriage very expensive, and as the Society does not pay this charge, artists do not like to undertake it on the mere chance of selection.'

A new, larger and much-improved building was ready in time for the Bristol Show of 1864. It combined the three essential qualities of efficient ventilation, perfect security from damp, and great portability. The exhibition's reputation was now firmly established and the Society felt entitled to point out to prospective exhibitors, especially painters, that 'a far greater number of persons visit the exhibition in one week than most provincial exhibitions attract during the three or four months they are open'. With the inducements of 'a spacious gallery, good light, immense attendance and quick sales', the Society felt that it should not be expected to show anything but pictures of good quality and that, in future, 'artists would not forward crude and hastily executed works, unworthy of their best powers'. There was more than a hint that Dorchester and Bristol were getting pictures that had no hope of a sale in London. A similar regrettable tendency might be noticed in the 'art manufacture division'. Here, instead of 'goods of the highest class', there was all too frequently to be seen 'a collection of productions such as is usually found in bazaars'. In 1870 the Secretary felt obliged to mention that the situation had deteriorated to such an extent that 'several of the exhibitors were persons who have their regular stalls at the Crystal Palace'.

By 1873 the position in the Fine Arts section had become serious. It was clear that 'artists of repute' were not fired with the same missionary fervour as the Bath & West Society itself. They were not being persuaded to send their best work for exhibition and possible sale at an agricultural show, however well attended. London, with the Royal Academy and numerous private galleries, gave them all the opportunities they needed.

A more realistic policy had therefore to be adopted. Good pictures would have to be obtained on loan from private collections. A determined effort was made to achieve this at Plymouth in 1873, the emphasis being on securing paintings by artists with West Country connections wherever possible.[32] The campaign was extremely successful. Seven paintings by Sir Joshua Reynolds, 'a native of Plympton, near Plymouth', were made available, four being lent by the Earl of Morley, from Saltram, and two by the Earl of Mount Edgcumbe. Other pictures were by Opie, Prout, Angelica Kauffman and Sir Thomas Lawrence, 'native of Bristol', whose 'finest work, "The Calmady Children", was kindly lent by its fortunate possessor, Mr Vincent Calmady'. The water-colours were 'all of the highest quality'. They included works by David Cox, De Wint, Pyne and William Cook, 'the last-named artist, a well-known and respected Plymothian, being represented by about 50 of his finest works'. With such pictures on view, it was hardly surprising that the Exhibition was 'visited by eager crowds, most of whom witnessed perhaps for the first time in their lives those grand works of English art of which this country is so justly proud'.

25. *Lustleigh (Devon)*, 1860 by Philip Mitchell, now in the City Museum and Art Gallery, Plymouth. Exhibited at the Plymouth Show, 1873.

It would be misleading to suggest, however, that all the pictures lent and shown at Plymouth were of quite this quality. *The Catalogue of Paintings, Drawings, Sketches, Etc.* at Plymouth in 1873 suggests that the watercolours reached a rather higher average than the oil paintings. A typical page of the watercolour section, showing the wide range of ownership and the concentration on local subjects, reads:

'310	Reubens (supposed)	Chalk Head	Lieut.-General Gascoyne
311	Payne, W.	Landing of George III, at Mt. Edgcumbe	H. Luscombe, Esq.
312	Payne, W.	Tinside, under the Hoe	H. Luscombe, Esq.
313	Cook, S. (the late)	Tintagel	Geo. Soltau, Sq.
314	Cook, S.	Bridge on Dartmoor	Geo. Soltau, Esq.
315	Payne, W.	Laurel Walk, Mount Edgcumbe	H. Luscombe, Esq.
316	Payne, W.	The Old Guildhall, Plymouth	Miss Hawker
317	Varley	Landscape	F. Hicks, Esq.
318	Dewint	Sketch	A. B. Bone, Esq. Junr.
319	Prout, S.	Tintagel Mill	W. Eastlake, Esq.
320	Cox, David	Dead Pheasant	Mrs Eberhardt
321	Cook, S.	Landscape—Autumn	Geo. D. Radford, Esq.
322	Prout, S.	Coast Scene	F. H. Goulding, Esq.
323	Calcott, Sir A., RA	On the Moor	J. W. L. Ashe, Esq.
324	Prout, S.	Launceston Castle	W. Eastlake, Esq.
325	Prout, S.	Rocky Point, between Plymouth and Stonehouse, where now the Great Western Docks are situated	Rev. J. M. Hawker
326	Wimperis, G. M.	Landscape	Geo. D. Radford, Esq.
327	Prout, S.	Worthing	Miss Norman
328	Cox, David	A Welsh Lake	Mrs Eberhardt
329	Jacobson, W.	Haldon	J. W. L. Ashe, Esq.
330	Jacobson, W.	River Scene	J. W. L. Ashe, Esq.
331	Giles, R. H.	Portrait of William Jacobson, Esq.	Mrs Jacobson
332	Cook, S.	South Pill, Saltash	A. Mudge, Esq.
333	Jacobson, W.	Series of Drawings	A. B. Bone, Esq., Junr.
334	Smith, Lieut.-Col. H.	Sire Gyserecht (Keyarts), Van Liefdale, tutor of Godfrey, the Third Duke of Brabant	Miss Smith
335	Prout, S.	Venice—Doge's Palace	Rev. J. M. Hawker
336	Cook, S.	Rough Tor and Brown Willy	Mrs Hicks
337	Cook, S.	Fowey Castle	Geo. D. Radford, Esq.
338	Cook, S.	View on the Tavy	Mrs Hicks
339	Cook, S.	Tintagel, Cornwall	Mark Grigg, Esq.
340	Cook, S.	The Wengern Alp	C. C. Whiteford, Esq.
341	Cook, S.	The Plym	George Soltau, Esq.
342	Cook, S.	Barn Pool, Mount Edgcumbe	H. Brown, Esq.
343	Cook, S.	The Wengern Alp	Mrs Oliver
344	Cook, S.	View of Plymouth	Lord Selborne
345	Cook, S.	Storm, North Wales	Mark Grigg, Esq.'

To assemble several hundred pictures from so many sources was no mean feat of organisation.

This system of persuading local people to lend pictures worked reasonably well for a number of years, although on occasions when the neighbourhood proved unproductive the Committee had to throw its net wider and enlist the support of 'those numerous possessors of pictures in London and elsewhere'. During the 1880s, however, greater use was made of the Society's Art Union, a mutual encouragement club for amateur artists. Members of this body submitted works for exhibition and sale at the Show, and awarded one another prizes, a habit not approved by the Society, because 'experience has long since shown that, being often persons ignorant of art, they are prone to select works of very inferior merit'. By employing independent judges, with more severe standards, the class of painting shown apparently improved.

The Society's Arts and Manufactures Committee continued to function until 1901, but without the energy and pioneering spirit that had characterised its work between 1860 and 1880. The exhibition that was got together each year represented an adventurous and far-sighted attempt to make English provincial life a little less provincial, at a time when the great majority of people in the rural areas had never seen a good picture in their lives, unless they happened to work as domestic servants in the kind of large household where there was something better than a series of pedestrian family portraits on the walls. We still have detailed descriptions of what was set out in the Art Building year by year, but what the tens of thousands of visitors thought of the Turners, the Reynolds and the majolica we shall, unfortunately, never know. As with any other attempt at mass education, the results were, inevitably, both unpredictable and uncontrollable.

But, during the 1850s and 1860s, the value of the Arts Section was never in doubt. In 1869 Mr Worsnop reported on what he had seen at the Southampton Show. The Bath & West Society, he said, 'is one of the great Art teachers of the day. It brings annually before a class of society, that cannot be reached in any other way, a varied and valuable collection of oil-paintings, water-colour drawings and works of ornamental art, seldom surpassed by exhibitions of a more stationary and lasting character; thereby awakening thoughts that otherwise might not have existed.'[33]

Largely as a result of Acland's enthusiasm and efforts, the Society sponsored or encouraged a number of schemes to improve the standard of rural education. The natural history exhibitions and competitions for school-children at the Annual Show were popular and much of the work was of a high standard. Acland, however, devoted most of his energy to what he called 'middle-class education',[34] a system of public examinations, with prizes, for boys educated in the West of England, mostly at grammar schools. The universities were involved, in itself a revolutionary step, and the first examination took place in Exeter in June 1857, with a distinguished panel of examiners, including several of Her Majesty's Inspectors, and Acland's own brother. The aim was to provide 'an examination which shall test the success of the education given, whether in schools or elsewhere, and thus at once give parents the power of discriminating efficient teachers, and teachers the opportunity of proving their own skill.'

Many of the boys—it was mid-Victorian England and so they were all boys—lived a long way from Exeter. The list of successful candidates at the first year's examinations, which took four days, included pupils from Bristol and Bath Grammar School, King Edward's School, Bath, and Kingswood School. Their ages ranged from 15 to 18 and they were the first children in Britain to receive an objective, outside assessment of their abilities. In a very short time, these examinations were being referred to as 'University Local Examinations', and in 1859 the Bath & West Society reported, not without pride, that its experiment had led to similar committees being set up in nine major cities in England.

One could continue for a long time to detail Acland's enthusiasms and ideas, his support for the Society's extremely successful Dairy Schools, the help and encouragement he gave to practical agricultural experiments of all kinds, many of them on his own land, his advocacy of a Department of Agriculture—but not a Minister of Agriculture, which he thought was un-British. After the death of his father in 1871, and his succession to the baronetcy, he had great personal responsibilities, with 15,000 acres in his charge. A wealthy man himself, he felt very keenly the need for a more equitable distribution of wealth in the countryside, and he spoke and wrote a great deal about this. In 1875, for instance, we have these reflections from him:

'When we think of the fearful difference between the enormous fortunes which have been handed down or which are being built up at the present day, and the poverty and misery which surround us in this country, the picture presented is truly alarming and almost heartbreaking. ... We have the fact that a large proportion of our population are ministering to the enjoyment of the richer classes, and it is a matter of great anxiety in the outlook for all statesmen and patriotic persons.'[35]

We can leave him, perhaps, with a penetrating assessment of his character, written in her diary, by his first wife's mother, Lady Mordaunt, shortly after his marriage. She sums him up in this way.

'Sense and ability and quickness of perception; fine temper; tender heart; generosity and disinterestedness; absence of variety; strong faith and warm charity. These qualities are of the highest order, and very active life would amend many of the defects. Dilatoriness and procrastination; indecision and too little self-confidence; undue importance given to little things; an eccentric dwelling on same, as for instance, his accounts; theoretical (very curious): a want of repose of character; scrupulousness approaching to superstition and arising from faults in education; he says self-indulgent; an inequality of spirits.'[36]

Acland resigned as Editor in 1859, and from then on found himself obliged to spend a great deal of time on his Parliamentary duties and on helping to run the family estates. It was, in one way, a good time to hand over his responsibilities. The Society was in a flourishing condition and its reputation had never been higher—the membership in 1859 had reached 1,258, the Prince Consort had just become a Member, the finances were very satisfactory and the policy of holding the Annual Meeting in a different town each year had fully justified itself. Acland could look back on the past ten years and feel that he had not wasted his time.

His greatest single achievement had been to build a bridge between scientists and ordinary practical farmers, and in this his most important ally had been someone who he himself was largely instrumental in getting appointed, Dr Augustus Voelcker, the Society's Consultant Chemist.

Born in Frankfurt in 1822, he had poor health as a child and was unable to attend school until he was twelve. At the age of sixteen he went to work as a pharmacist's assistant, first at Frankfurt and then at Schaffhausen. When he was twenty-two he managed to get to the University of Göttingen to study chemistry, and somehow to finance himself there for the customary four years. After graduating in 1846 he left Germany and went to Utrecht, to work as principal assistant to Professor Mulder. Mulder specialised in physiological chemistry, especially its relation to horticulture and agriculture. Voelcker spent less than a year in Utrecht, but his work there with Mulder settled the direction of his future research and the field in which he was to earn his living.

In 1847 he managed, with Mulder's help, to get a post in Edinburgh as assistant to James Finlay Weir Johnson. Johnson happened to be chemist to the Agricultural Chemistry Association of Scotland, and this gave Voelcker his chance to come into close contact with farmers and their problems, and, incidentally, to improve his English. Johnson was an exceedingly busy man, much in demand, and he sub-contracted his lectures at Durham University to Voelcker. Despite his rather curious English, Voelcker was apparently successful at Durham. While he was there he had the good fortune to get on friendly terms with another visiting lecturer, George Wilson, the Regius Professor of Technology at Edinburgh. Wilson made Voelcker realise quite clearly that his future lay in applied research and after two years in Edinburgh, he came south, to become the first Professor of Chemistry at the Royal Agricultural College at Cirencester.

Here he had a laboratory of his own, with technical assistance and £400 a year. Men of Voelcker's calibre, specialising in the applications of chemistry to agriculture, were rare and, in 1854, the Bath & West Society reckoned, quite rightly, that it was lucky to have been able to persuade him to become its first Consultant Chemist. The new appointment was explained to members in this way: 'A profound acquaintance with chemistry, which is daily becoming more extensive and complicated, an acquaintance with the requirements of Farming as practised in England, and also in the habits of Farmers, a considerable amount of practical common sense, and especially a position of independence in reference to all connections with trade, are qualifications not often combined in the same person.'[37]

Voelcker received no honorarium from the Society.[38] His financial reward came entirely in the form of fees paid by farmers to have samples of soil, water, feeding-stuffs and fertilisers analysed. Since much of the routine work in connection with this would have been carried out by Voelcker's laboratory assistants at Cirencester, or in some cases by students, the burden on him personally was probably not too onerous. The fees charged ranged from 7s 6d to 5 guineas, according to the substance to be analysed. Payment had to be made in advance and samples could either be posted to him direct, at Cirencester, or left for forwarding at one of six collecting centres within the Society's area.

There was much point in the emphasis on Voelcker being in 'a position of independence in reference to all connections with trade', because for some years his most important advisory work for the Society consisted of analysing and reporting on samples of fertilisers, in an attempt to show farmers, who were often of almost childlike credulity, how to protect themselves against the swindling which was so widespread and scandalous within the trade. The Society itself, of course, was much too gentlemanly an organisation to use crude words like 'swindling'. It preferred to talk of 'the extensive and serious losses traceable to the adulteration of manures'. But everyone understood what was meant. In order to do something about the situation, farmers had to be taught what the constituents of fertilisers were, or should be, and how to measure value for money, but, as the Society said, 'it requires peculiar qualities to enable any one to render this information accessible to the generality of farmers'. Voelcker had these 'peculiar qualities'. He was a first-class populariser and, unlike so many of his fellow scientists, used straightforward language, both in his lectures and his articles.

There are several photographs of him. They all show a kindly, humorous, gentle face, sufficiently like the Victorian stereotype of a scientist to impress the man in the street, but somehow with more of the German musician than the German researcher about him. One can easily understand that audiences took to him readily. But his success as a speaker had to be worked for. By the time he arrived at Cirencester, he had come to realise that his accent and his markedly German way of putting sentences together were preventing him from being really satisfactory as a lecturer and a writer. He took drastic steps to improve. To become more

26. Augustus Voelcker. From the portrait in the possession of the Royal
Agricultural College, Cirencester.

thoroughly English, he gave up speaking his native German altogether, even with
his own family—he had a German wife—and he worked very hard to sound at
least as English as Prince Albert did. He later published articles which contain little trace of foreign idioms or constructions, although he may, of course, have
received a little editorial assistance from time to time.

He developed a notable talent for getting on good terms with audiences of
English farmers. For a number of years during the Fifties and early Sixties, he
was in the habit of giving lectures in towns up and down the West Country. Time
was his most precious commodity and travel had to be cut to a minimum. His

system was to give a lecture every evening, apart from Sunday, for a week or a fortnight. In one year, for instance, he was in Plymouth on Monday, to talk about *Superphosphate*, on Tuesday in Taunton, with *Clay Soils* and on Wednesday in Bath, where the subject was *The Connection of Chemistry with the Practice of Farming*. The lectures and discussions were fully reported in the local newspapers, with long passages verbatim, and it is consequently not too difficult to get an impression of his style. He had a direct, lucid way of putting his points across, and there is no doubt that he was a very successful speaker. In one of the first papers he wrote for publication by the Society, 'On the Agricultural and Commercial Value of Some Artificial Manures, and on their Adulteration', he warns that 'If there ever was a time when the agriculturist had need to exercise especial caution in purchasing artificial manures, that time is the present; for the practice of adulterating standard artificial fertilizers, such as guano, superphosphate of lime, nitrate of soda, etc., has reached an alarming point. . . . Some artificial manures are actually sold and bought at double or triple the price which they are worth.'[39]

He gives examples, supported by detailed chemical analysis, of the kind of frauds that were being perpetrated and tells farmers how to defend themselves. 'We would observe that the presentation of a chemical analysis by the dealer is in itself no guarantee of the genuineness or value of a manure. What we would recommend to the purchaser is, to demand an analysis of the dealer, and, in case he should not understand the meaning of chemical terms, to have the analysis explained by somebody who does. If satisfactory, he should order the manure according to the furnished analysis. On delivery of the manure, a sample from the bulk should be selected for analysis. A comparison of this analysis with that furnished by the dealer then will show whether or not the contract has been fulfilled, and measures can be taken accordingly.'[40]

The combined effect of his analyses and his lectures produced a marked change in the situation within two years. 'We no longer hear,' the Society reported, 'of adulterations under the name of "silicious constituents". The fraudulent dealer is disappearing or hiding in obscure corners, well knowing that the members of our Society have an infallible test at hand.'[41] With Dr Voelcker as their watchdog, no member of the Bath & West Society need have been in the position of the unfortunate Archdeacon in Trollope's *Barchester Towers*: 'I got a ton and a half of guano at Bradley's in High Street, and it was a complete take-in. I don't believe there was five hundredweight of guano in it.'

Voelcker's work covered a very wide range, 50-page articles are frequent, and one wonders very much how, in those days before typewriters, he found time to cope with them, in the midst of an exceptionally busy life. His published reports for the Royal Agricultural Society and the Bath & West Society alone would have made half-a-dozen substantial volumes, although he never published a book as such. One of his particular interests was cheese-making. He was a pioneer in the application of scientific principles to this ancient craft, in which success was supposed to be a matter of luck or intuition and failure merely an act of God, and poured scorn on the popular idea that there was something inexplicable and mysterious about the techniques and processes involved. 'All that is mysterious about it,' he wrote, 'is purely accidental. The process in itself is very simple, and accords well with scientific principles so far as these have been ascertained; but skilful management is perhaps rather the exception than the rule.'[42]

Voelcker provided the indispensable theoretical knowledge on which one of the Society's greatest Victorian achievements, its pioneering cheesemaking schools, were based. He acted as an invaluable bridge between the laboratory and the dairy. Consider, for example, the way he writes about thermometers. These 'invaluable aids to scientific cheese-making,' he wrote, 'are seldom in use. Even when they are hung up in the dairy, they are more frequently regarded as curious but useless ornaments than trustworthy guides, and therefore are seldom put into requisition. In fact, most dairymaids are guided entirely by their own feelings; and as these are as variable as those of other mortals, the temperature of the milk when it is 'set' (that is, when the rennet is added) is often either too high or too low. They mostly profess to know the temperature of the milk to a nicety, and feel almost insulted if you tell them that much less reliance can be placed on the indication of ever so experienced a hand upon an instrument which contracts and expands according to a fixed law, uninfluenced by the many disturbing causes to which a living body is necessarily subject.

'It is really amusing to see the animosity with which some people look upon the thermometer. It is true there are not many dairies in which it may not be found; but if we took pains to ascertain in how many of these it is in constant use, I believe that the proportion would not exceed 5 per cent.'[43]

It is interesting to notice that a long article on cheese-making, published in the 1857 volume of the *Journal* contains no reference to thermometers or to any other instruments. This article, 'Report on Cheese-Making by the Deputation sent by the Ayrshire Agricultural Association, to inquire into the Methods of making Cheese in the Counties of Gloucester, Wilts, and Somerset', describes visits to three representative farms and to a number of cheese-factors. The dairies showed 'nice, careful management', and 'they all seemed to be models of cleanliness and good order. The condition of the floors, the walls, and particularly the utensils, is carefully attended to; the servants are dressed tidily, and in a manner suitable for their work; and the whole management has that attractive appearance which feminine neatness and good taste, combined with intelligence, are sure to provide.'

At Mr Leonard's farm, 'Water End, near Frocester, in the Vale of Berkeley, when the vats are all filled, they are reversed, and the bottom one placed uppermost. The top cheese is taken, and a triangular paring, about an inch broad at the base, is cut off round the edge. It is then turned into a whey-cloth, the vat is rinsed with a little whey, and the cheese is put into it with the cloth under. The edge that is now uppermost is pared round as the other had been, and a portion of the curd, in the form of an inverted cone, is cut out of the centre of the cheese. This is called "cutting out the witch", and we have been informed that the practice is seldom omitted by a Berkeley dairymaid. Along with an old horse-shoe over the door it forms a perfectly sufficient safeguard against witchcraft.'

Voelcker certainly did not believe that 'cleanliness and good order' were in themselves sufficient to ensure the production of good cheese, and he would have recommended a thermometer (no effective and simple acidometer yet existed) in preference to a horse-shoe. But he understood the human barriers to scientific progress and he knew very well that patience was required, if science was to make its way.

In 1863 he resigned his post at Cirencester—the College was being reorganised and there were policy changes—and set up in private practice in London as an analytical chemist. The business built up quickly and is still in existence, in Tudor Street, under the name of Dr Augustus Voelcker and Sons, Ltd. It is still run by Voelckers, with members of the third and fourth generations actively maintaining its traditions. A chemist may breed a chemist once but for the line to continue unbroken from father to great-grandson, and in the same building, is surely unusual, and quite possibly unique.

27. Frederick James Lloyd, the Society's Consultant Chemist.

Augustus Voelcker continued his association with the Bath & West Society after moving to London, where he combined private practice with an appointment as Professor of Chemistry at the Royal Agricultural Society. At the R.A.S. he succeeded the Society's first chemist, J. T. Way. There was considerable prestige attached to the post, but not a great deal of money—£500, compared with the £400 he received at Cirencester. To maintain himself and his family, and to be able to continue his own research, the private practice was essential. He had a formidable capacity for work—his research was carried on in his spare time—and a number of the next generation of English agricultural chemists, including F. J. Lloyd, Bernard Dyer, and his own sons were trained in his laboratory.

During the whole of Voelcker's working lifetime, Britain treated her agricultural chemists, like most of her scientists and technologists, in a foolishly parsimonous way. It was the lack of men who were in a financial position to devote anything like their whole time to research that accounts for the comparatively small number of British discoveries in this field during the middle part of the century. Voelcker's efforts and influence in maintaining the scientific approach to agriculture was both important and heroic. His election to the Royal Society in 1870 was thoroughly well deserved.

Dr Voelcker's great grandson is now the Bath & West Society's Honorary Consulting Chemist.

An article published in 1861 by the Bath & West Society in its Journal *listed the main developments in agriculture during the previous twenty years. At the head of the list was land-drainage, to which Parliament had voted large subsidies. Drainage and land-improvement companies had been set up, and 'one of these, belonging to the district of the Bath & West of England Society, though not confining its operations there—the West of England and South Wales Land Drainage Company—established in 1844, has since accomplished the improvement in this and other ways of thousands of acres of land, at an expenditure of more than 400,000£; and other similar associations have been even more extensively engaged; so that Mr Denton, one of our leading authorities on land drainage, estimates the expenditure between 1847 and 1860 on this one means of agricultural improvement alone, at 5,250,000£.' The West of England Company was formed by leading West Country landowners, like Sir Thomas Dyke Acland, Lord Clinton and Sir John Kennaway. Its incorporating Act allowed the Company to advance money to landowners on the security of their estates.*

Guano and artificial fertilisers were much more widely used, wheel-ploughs had become almost universal, and so had turnip-cutters, oilcake-breakers and root-pulpers, resulting in 'the economical use of much food that has hitherto been wasted.'

Cultivators were now commonly seen, largely as a result of land drainage. 'Stirring implements,' Mr Morton pointed out, 'cannot be used when land is wet.' On heavy land especially, spring cultivations had taken the place of a second ploughing. This change in practice had to be considered in relation to the growth of the big machinery manufacturers, such as Ransomes and Sims, R. Hornsby & Son, and James & Frederick Howard, who were providing the new equipment farmers needed and demonstrating it at the local shows.

The mobile threshing machine had been introduced in 1841. There was, in Mr Morton's view, 'no class of machines of which the usefulness and economy have so rapidly commended themselves to farmers generally. Of this the enormous extension of the locomotive steam-engine till lately used exclusively for the driving of the moveable thrashing-machine, is sufficient proof. At the close of 1859, returns from the leading manufacturers enabled me to state that within the past four years 40,000 horse-power had been added to the forces used in agriculture in steam alone. Messrs. Clayton and Shuttleworth of Lincoln, Garrett of Saxmundham, Hornsby of Grantham, Ransome of Ipswich, and Tuxford of Boston, were then alone furnishing 10,000 horse-power annually to

*the farmer; and Messrs. Clayton, of Lincoln, were sending out ten engines week-
ly, or 4,000 horse-power per annum.'*

*The reaping machine had been invented many years ago in Scotland, but 'it
was not until after its appearance in the American Department of the Great
Exhibition in 1851, that it was first introduced to English agriculture; and now it
bids fair to make almost as great a change in the summer work of the
agricultural labourer here, as the thrashing-machine has effected in his labour
during winter.'*

*It needs to be remembered, however, that farm labourers earned a good deal
more in the north than they did in the south, and that this differential persisted
throughout the century. If the average weekly wage in 1850–51 for the whole of
England is taken as 100, the movement over the next twenty years was:*[1]

	1850–51	1869–70	1872
Northern Counties	130	165	188
Midland Counties	104	138	161
Eastern Counties	84	114	138
South and South-Western Counties	83	111	131

*From the 1850s onwards, the lot of the farm worker improved markedly, both
in wages and in better living conditions and social amenities. Without these improvements, farmers would have been unable to hold sufficient labour to staff
their new systems of mixed agriculture. Whether, from a landowner's point of
view, improvements contributed very much to profits is another matter.*[2] *The
evidence from the estates of the Marquis of Bath, the Earl of Pembroke, the Duke
of Bedford and the Duke of Northumberland, all of whom spent a great deal of
money on improvements during the middle part of the century, suggests that a
2–3 per cent return on the investment was all that could be expected, even with increased rents. It has been suggested that 'much of the landowner's investment in
high farming never was an economic proposition'.*[3]

Appealing to a wider public

The last volume of the *Journal* to be edited by Acland[4] includes an article of an all-too-rare type, a practical farmer describing how he went about his business. 'A Report of Personal Experience on a Farm, upwards of 500 acres, situate at Kilmington, in the County of Somerset. By Joseph Lush, of Brewham House, near Bruton' is simple, direct and very refreshing to read, and one can easily understand why the Editor decided to print it.

'None of my land requires draining.

'I have no orchard. I cannot grow apple-trees in my garden by any means that have yet been tried.

'I cannot say much of the manufacture of dung, and I think the less that is done in that way in the yards the better. I endeavour to mix the different qualities as well as I can, and get it into the soil with as little delay as possible. From the time I finish wheat sowing, which is in November, to the time I begin sowing in the following October, I take every opportunity of clearing the yards, and conveying it to the fields for clover, vetches, rape, and turnips as a preparation for wheat, and I particularly desire that not an atom or drop of the excrements of any animal be allowed to drop in the yards or houses, without straw being under them to receive and absorb it.

'As to the treatment of labourers, I think I stand tolerably well with them, as I seldom change. I act on the commercial principle of buying in the cheapest market, but do not consider the lowest price always the cheapest. I hire mostly by the week, but my reaping, mowing, hedging, and hoeing is done by measure, and some other work also, as occasion requires. I have for upwards of twenty-five years paid my labourers weekly on Friday evening, in cash separately, so that none should have an excuse to go to the ale-house or shop for change; and although the whole time I had business transactions with the landlord of the inn in the village, and received large sums in each year from him, I never sent a labourer there for wages. During the whole time of my improvements, I superintended nearly always on foot, and assisted in every operation that took place on the farm, and measured the whole of my piece-work (with very few exceptions) myself, and very seldom had a complaint of any mistake; in fact, the men generally preferred my measuring their work, as they then saved their moiety of the expense; and although I have had as many, on one or two occasions, as one hundred persons in my employ at one time, I have never had a magistrate's summons for any one in my service, nor was I ever summoned by one of them on any matter whatever.'

This, one feels, is the authentic voice of the man who cared deeply about his land and about the men he employed to work on it, who woke up early and went to bed tired, had little time for reading and thought for a very long time before changing his methods. An article such as the one we have quoted would surely have taken him a long time to write, and he must have taken a good deal of persuading before he decided to set pen to paper. Acland was the type of editor who managed, against all the odds, to make the solid, practical man vocal, just occasionally.

The editor who succeeded him was a very different character. Born in 1817, in Scarborough, Josiah Goodwin was a professional journalist. His career has been sketched by a later Secretary, who knew him well, Thomas Plowman.[5] 'He was,' Plowman wrote, 'an exceedingly expert shorthand writer, and his intelligence and accuracy as a note-taker led to his being engaged to furnish official reports of many important arbitration cases in London and the Midlands arising out of the first development of railway enterprise. He subsequently obtained a wide experience of press work generally, and the "Birmingham Advertiser", and "Wilts County Mirror", and the "Exeter Gazette" all came under his editorship.

'In 1859 Mr Goodwin first became connected with the Bath & West of England Society, having been appointed Editor of its *Journal* and, on the retirement of the late Mr H. St J. Maule, in 1866, he succeeded him as Secretary, holding the two offices conjointly. During his tenure of the last-named office the Society's area of usefulness was much extended, and there were new and important developments in its operations. In 1868 he added to his responsibilities by taking temporary charge of the Journal of the Royal Agricultural Society of England, and he brought out two of its numbers during the interregnum which occurred between the death of Mr Frere and the appointment of a new Editor, the late Mr H. M. Jenkins.'

Josiah Goodwin was evidently a sociable man. 'His conversational powers,' Plowman recalled, 'afforded much pleasure to those who enjoyed his friendship, for his career had furnished him with many entertaining and instructive reminiscences, the interest of which was enhanced by the telling. He had not only seen much of men and things under many different aspects, but his memory could be relied on to honour the drafts he drew upon it. He had at all times a certain stateliness of speech and a carefulness of expression suggestive of a period when life was less of a whirl and rush than now, and when men took time to weigh their words before they uttered them. His genial courtesy—which was so natural to him—and kindly consideration for the feelings of others, rendered intercourse with him a pleasant experience at all times, as those who were brought into personal contact with him, and especially his colleagues, could testify.'

For the last years of his life Goodwin was crippled with rheumatism, a disability which compelled him to give up the Secretaryship in 1882, but he continued to edit the *Journal* until the year of his death, in 1889.

Immediately he was engaged as Editor—it was, of course, not a full-time job—Goodwin had to implement a new policy for publishing the *Journal*, in order that topical material should be still topical when it reached the hands of Members. In 1860[6] it was announced that 'the desire expressed by members of the Society that the *Journal* should be published as soon as possible after the yearly Exhibition of Stock, Implements, etc. has induced the Committee to alter the plan of publication. Instead of reserving the whole of the available materials for one bulky volume at the close of the Society's year, they purpose hereafter, in accordance with the practice of the Royal Agricultural Society, to issue the *Journal* in two parts: the first to appear when practicable in the autumn, the second in the spring.'

The system of offering prizes for essays printed in the *Journal* was attracting very few entries. 'It would seem,' said the Committee,[7] 'that the competitive system, involving the risk of failure after long and persevering effort, has little attraction for experienced writers. This subject has not escaped the attention of the Council: though disclaiming all pretensions to novelty for its own sake, they expect money's worth in return for their outlay. To meet the difficulty, several plans have been suggested: the most SATISFACTORY would be, that practical men in the West of England should communicate the results of their experiences to the Editor, and thus render the Journal a repertory of applied science and practical knowledge.

'The Journal Committee are authorised by the Council to engage literary assistance to a limited extent, and will therefore be happy to enter into communication with any author willing to contribute to the Journal.'

What this meant, in plain terms, was that those 'practical men in the West of England' who had useful material to contribute, but who were lacking in literary skill, could have their rough-and-ready articles polished up for them by rewrite men, hired and paid for by the Society, which now had a little money to spare for the purpose. But some people, whose work needed no polishing, were indefatigable writers of prize essays. Chief among them was Henry Tanner, Professor of Agriculture and Rural Economy at Queen's College, Birmingham. He had three essays published in the 1860 volume alone.

The 1861 *Journal* has an article, 'Recent Improvements in Dairy Practice', by Joseph Handing, of Marksbury, Somerset, which was reprinted from the *Journal of the Royal Agricultural Society*. In it he pleads for a scientific method of testing acidity during cheese-making. 'The instrument which we want,' he wrote, 'is one which will show us the exact amount of acid present, that we may know when to introduce the rennet, and in what quantity. It is true, we have litmus-paper, but this only indicates the presence of acid without measuring the *quantity* present. Whilst searching for such an instrument as this among opticians and chemists for several years past, I have been recommended to try one or two chemical methods, the best of which is by Dr Cameron, of Dublin. None of these tests, however, are sufficiently simple to be of much use to the practical dairywoman, who wants an instrument effective and simple, by which she can as easily test the amount of acid present, as she can by the thermometer ascertain the degree of heat.'

The Society's Consulting Chemist, Dr Voelcker, strongly disagreed with this, in another article originally published by the Royal Agricultural Society.[8] 'A great deal has been said and written,' he noted, 'with respect to the great utility to the dairyman of an instrument by means of which the amount of acid in sour milk might be accurately and readily determined. A careful study of the action of rennet on milk, however, has led me to the conclusion that the more carefully milk is prevented from getting sour, and, consequently, the less opportunity there is for the use of an acidometer, the more likely the cheese is to turn out good. Indeed, the acidometer appears to me a useless instrument—a scientific toy which can never be turned to any practical account. If by accident the milk has become sour, the fact soon manifests itself sufficiently to the taste. An experienced dairymaid will even form a tolerably good opinion of the relative proportions of acid in the milk on different days and arrange her proceedings accordingly. Moreover, the knowledge of the precise amount of acid in the milk does not help us much. When milk has turned sour, the best thing to do is to hasten on the process of cheese-making as much as possible.'

Matters such as this were of particular importance in the South-West, where dairy-farming was becoming increasingly important each year. As Professor Tanner had put it, in one of his prize essays, 'There is no department of agriculture which holds out greater inducements for the exercise of judgment and the employment of capital than farming grass-land. In the West of England it may become the most profitable branch of agricultural industry, and it also offers a more certain return for capital than the tillage of the soil. Take them for your motto, "Plough less and graze more".'[9]

Sheep, too, contributed a great deal to the income of many farms in the western counties, and the 1861 volume has an authoritative article on one of the major nineteenth century plagues, 'On the Disease of Sheep, commonly known as Rot, Coathe, or Bane', by G. T. Brown, Veterinary Professor at the Royal Agricultural College, Cirencester.[10] This article is especially notable for its illustrations, made by microscope enlargement, of such things as the gall-bladder of a sheep affected by rot.

Foot-and-mouth disease or, as it was then known, the cattle plague, was an unceasing worry to nineteenth century farmers, as it is, indeed, today. There was a serious outbreak in 1865 and in August the Society held a special meeting of the Council in order to decide on the best policy to follow. A Committee was appointed 'to collect and disseminate information', and it circulated a document to farmers and local authorities throughout the South-West, describing the symptoms of the disease and indicating the precautions which should be taken. At this time, there were no Government regulations dealing with the matter, so that private bodies like the Bath & West Society carried a heavy burden of responsibility for any recommendations they might make. At a Council meeting in December, when the outbreak showed no signs of abating, these resolutions were passed:

'1. That it is the opinion of the Council of the Bath & West of England Society for the Encouragement of Agriculture, Arts, Manufactures, and Commerce, that an effectual means of suppressing the Cattle Plague would be to stop for a time all fairs and markets, and collections of cattle for the purposes of exhibition or sale.

2. That all foreign cattle, sheep, and pigs should be slaughtered at the port of disembarkation, and their skins disinfected forthwith.

3. That no cattle, sheep, or pigs be allowed to travel upon any public road from any farm or place upon, or in, which there is, or has been, within two months, any case of Cattle Plague.

4. That the necessary orders for carrying out these resolutions be issued by the Government, and that the Local Authorities be required to enforce them, so that the practice may, as far as possible, be uniform throughout the country.

5. That whereas by the adoption of these measures, the trade in store cattle will be practically suspended, the Council of the Bath & West of England Society recommend that these resolutions shall not remain in force beyond the 1st of March, unless the Government should at that time consider it necessary to continue them for a further limited period.'[11]

Under Goodwin's expert and not ill-rewarded guidance—by 1863 he was being paid £105 a year, £5 more than the Secretary—the *Journal* became a publication of real importance, which reflected the full range of activities in the agricultural field. For those with the time to spare, these volumes of the 1860s and 1870s make excellent reading. One can dip into any one of them and discover a great deal of interesting information, well written-up and presented. An anthology of extracts would make a very pleasant book in its own right.

Here we can do no more than offer brief illustrations of what was achieved, hoping that, with their appetite whetted, readers will feel drawn to these old monuments to enthusiasm and progress for themselves.

There are, for example, some fine articles on machinery. At the Southampton Show in 1869, 'the department of "Machinery in Motion" was on quite a royal scale, comprising no fewer than fifty-four stands; while fifty-eight steam-engines formed an imposing line of funnels, fly-wheels, and driving-belts, such as can be witnessed in no other Show-yard but that of the Bath & West of England Society

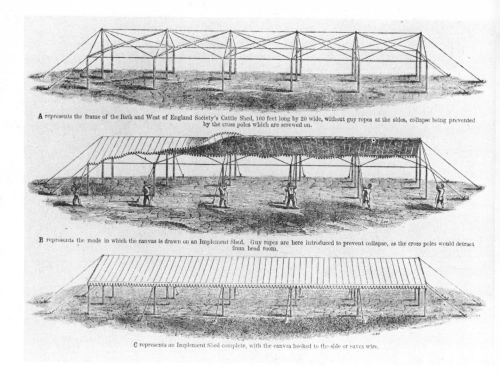

A represents the frame of the Bath and West of England Society's Cattle Shed, 100 feet long by 20 wide, without guy ropes at the sides, collapse being prevented by the cross poles which are screwed on.

B represents the mode in which the canvas is drawn on an Implement Shed. Guy ropes are here introduced to prevent collapse, as the cross poles would detract from head room.

C represents an Implement Shed complete, with the canvas hooked to the side or eaves wire.

28. Bath & West cattle and implement sheds, 1862.

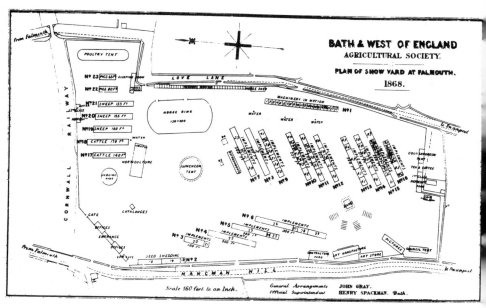

29. Plan of Showyard at Falmouth, 1868.

where the arrangements are always happily planned with a view to artistic effect. At the head of this extraordinary array of engines and machines stood Messrs. Clayton and Shuttleworth's machinery; and we were informed that, great as the show appeared, it fell far short of what that one firm at Lincoln turn out in a single month; for, in May last, they manufactured and sent away no less than one hundred and two steam-engines and as many thrashing-machines, the majority going to foreign countries. Twelve hundred men are employed at the works, but the immense production is accounted for by the fact that mechanism is used for almost every process in construction; even the finishing of small parts—such as glands, eccentric straps, and so on, is done by machine, and the engines exhibited at Southampton had scarcely had a hand tool upon them.'

The machinery section had a number of inventions by Mr J. Grant, of London. These included 'a very clever little contrivance, called the "automaton music leaf-turner", applicable to music-stands or to pianofortes, and most easily and conveniently fulfilling its duty at a touch of the performer's foot'.[12]

The machinery section at Hereford in 1876, included a number of American implements, including potato planters and diggers, and two delightful pieces of equipment:

'The Patent Horse and Cattle Groomer', shown by Newton Wilson and Co., of London, is a novel contrivance for grooming and cleaning horses and horned cattle. It may be worked by hand- or steam-power, the motor setting in operation a revolving brush worked from universal joints; this brush is applied to the hair and skin of the animal, much after the manner of a hairdresser's patent rotary "head cleaner", as used on the human biped, and with much the same result. The value of such a machine would only be apparent in establishments where a large stud of horses is kept. The horses cleaned by this machine seem to enjoy the operation, and the skin is certainly thoroughly cleansed and made to wear a sleek and glossy appearance.

'Brookes and Co., of Birmingham, exhibited their "Patent Lac Trephoer", an invention likely to be approved—with the exception, perhaps, of its queer Latin-Greek name—by agriculturists and breeders generally. It is a curious and useful machine in the nature of an artificial mother, for rearing calves, pigs, lambs, colts, puppies, and all kinds of young animals. A large percentage of young stock is annually lost to the country for want of attention and nourishment through the critical period of suckling, owing to the loss of the dam, or inability on her part to furnish the natural sustenance for her progeny. Brookes's apparatus supplies this in a simple and efficient manner. This mechanical "wet nurse" consists of a metallic case, easily cleansed, containing a tin reservoir for the reception of the liquid, with a number of tubes issuing from the outside, and provided with mouthpieces, in which are placed artificial teats of india-rubber. The animals are in this way fed by the natural act of sucking, thereby exciting the glands to healthy action, and allowing the flow of saliva to become mingled with their food in the most perfectly natural way. For calves and lambs the machine is especially useful, and the price is so moderate as to be scarcely worth consideration.'[13]

At the 1878 Show at Oxford there was a large exhibition of carriages, with wide ranges of accessories. Among these was Marston's new Patent Communicator, 'a very simple and ingenious contrivance for directing the coachman, likely to be a great boon to professional men using cabs and broughams. In this arrangement there are two dials with fingers and a bell; one dial is fixed inside, and the other outside the vehicle, opposite the passenger and driver, and connected with a direct spindle. The dials have printed on them all the instructions requisite for directing the driver—such as "stop", "faster", "slower", "turn to the right", "turn to the left", "pull up to the right", etc.; and as the finger inside is moved it moves the outer fingers, and at the same time strikes a bell which signals the driver's attention to the dial, where his employer's order is plain before him. This apparatus cannot be improperly used on the carriage, and it is so extremely simple that it cannot get out of order.'[14]

In 1879, the American reaper-binders aroused great interest. 'The impression received from witnessing the work of the machines is that the great expense of manual labour in tying corn may be dispensed with; that the tying may be done more rapidly, and, what is of far greater importance, the harvesting may be accomplished with fewer hands. Whether wire or string or even straw itself, will be the most generally used binding material the Exeter trials did not foreshadow; none of the string binders now before the public having been present at the Meeting.'[15]

The Society was proud of the way its machinery section was arranged. It disapproved of the scheme found at a number of other Shows, which put all the ploughs together, irrespective of makers, all the harrows together, and so on. 'This system,' it felt, 'however instructive to the public and handy for rapid comparison, is intolerable to exhibitors; ruining their chance of setting-out a fine well-filled stand, and hindering business, from their attention being required in several scattered parts of the ground. The Bath & West of England Society wisely limit the useful classification to the reference index of their Catalogue; by consulting which any visitor can walk straight to the stand of any particular exhibits wished for, or can direct his steps in succession to all the carts, all the threshing-machines, all the mills, or whatever kind of apparatus he may be most interested in.'[16]

The *Journal* devoted a great deal of attention to the dairying industry, which was developing rapidly, mainly in order to supply the profitable London market. The first wholesale milk depot to be established to meet the demands of London was opened at Semley, in Dorset, by the side of the main Exeter–Waterloo railway line. It is still operating. Two years later the Anglo-Swiss Company (now Nestlés) opened Wiltshire's pioneering milk processing factory, at Chippenham, and in the early 1880s a local farmer and businessman, Charles Maggs, set up a milk-collecting depot and butter factory at Melksham—the genesis of United Dairies, which later became even bigger, as Unigate.

A paper published in the *Journal* in 1870, by 'W.B.', drew attention to the problem of transporting milk by rail. 'The great drawbacks to the profitable production and sale of country milk experienced hitherto,' he wrote, 'have been the slow and objectionable modes of conveyance, and the limited demand of towns—town milk entering successfully into competition with country milk, both as to price and quality. Watery, thin, and inferior, as much of our town milk to this day is, it finds a more ready sale, taking the different seasons over, than the milk that comes from the country, and much of this preference given to the former arises from the quality of the latter being injured by the imperfect modes of conveyance in use. Of late years railways have done much to improve the commerce of country milk; but in the extremes of cold during winter, and heat during summer, and when the electricity of the atmosphere is in a greatly excited state, it suffers in such a manner as to disappoint town dairymen who rely upon their supplies from the country, and these mishaps are in favour of town milk. The trade in country milk, nevertheless, continues gradually to increase. Necessity, with her inventions, having come to the rescue of the producer with numerous improvements, so as to obviate the mishaps in question, and thus, less or more, turn the balance the other way. There is, however, yet much need for further improvements in the country milk trade.

'The erection of butter and cheese factories will, doubtless, give a new stimulus to the profitable production and trade in country milk. They will do so, partly from their closer proximity to the farms of producers, but chiefly from the more improved appliances, chemical and mechanical, which they will bring to bear upon the labours of sellers and buyers, the trade being thus better organised. And more than this, for these factories as they spring up will prove so many new markets for milk, consequently they will improve, not only their own trade with their respective customers, but also the country trade generally with towns. They will further encourage individual farmers or farming companies (as sewage farming companies) to produce milk, and connect it with butter and cheese, on the factory system, in places where the milk cannot be sold to towns. From these preliminary observations it will be seen that no second sight is required to perceive that dairy farming, as it is technically termed, is to form an interesting and popular subject for some time to come.'

W.B. emphasised that, as every farmer knew, from farm to railway, on the railway, and from railway to consumer, milk had a rough ride. The railway companies, he felt, should improve the springing of their waggons and consider raising the temperature of the milk-waggons above souring point by using waste steam from the engine. At this date, of course, the milk was not pasteurised at the collecting depot.

Another article[17] by Morgan Evans and William Eassie, explained that the increased demand for butter and cheese gave the British farmer an enormous opportunity, provided he realised what the market needed and invested in proper equipment. 'At no time,' they believed, 'has there been a greater demand for good sound butter and cheese than at present. The people consume better food now than they did twenty years ago, and partake of it more frequently. The high price of animal food of late years has been considerably affected by the improved diet of our lower classes, whose tastes are fast becoming almost epicurean. But animal food has nearly reached prohibitive prices. As the price of meat advances the demand for butter and cheese increases. The instinct and tastes of the people lead them to compensate for deficiency at the butcher's stall by turning to the butterman and cheesemonger. And of all substitutes for meat there is, probably, none better than cheese. Its nutritive qualities are equal to beef and mutton, and it is cheaper than either. We may safely anticipate that the present high price of meat will lead the working classes into a more just appreciation of cheese as a valuable food, and also to a more precise understanding of the merits of butter as a substitute for animal fats in supporting respiration and giving heat to the human frame. But the success of butter and cheese as articles of consumption will, after all, depend more on the public palate than on the public mind. Few people regulate their diet in accordance with a theory. They eat and drink what they like best. But when the taste is pleased, and the constitution satisfied, they are not slow to learn the merits of what they eat, as to cheapness and nutritive properties, and they choose accordingly.

'The present very large consumption of cheese in England is, doubtless, in a great degree due to a general improvement in its quality. A good cheese was a rarity in the time of our grandfathers, and when they found one they did not begrudge pouring into it a bottle of the best old port. There is now an abundance of good cheese in the market everywhere, as like in shape as two peas, and as like in taste as two cheeses from the same vat and from under the same press. This is the result of uniform processes in the making, the aid of science, and skilful management. Cheese is not now manufactured anyway by accident. It is made by design, just as much as the watch Paley imagined some one may pick up on a common, and on which supposed circumstance the eminent theologian based his celebrated argument. Anyone who found a cheese on Salisbury Plain in the present day, if he never saw a cheese before, might swear from its beautiful rotundity and delicious flavour that a human hand and human brain had been at the making of it, and that it did not grow there. The old-fashioned awkward make, as tough as shoe leather, still found in many districts, might in the same place be mistaken for an excrescence of the soil—a natural inedible fungus whose origin was obscure, and whose destiny was a mystery.

'At the present crisis, then, it becomes important that our butter and cheese should be made of a quality to suit the fastidious tastes of our people, and be at a price which, while being remunerative to the maker, shall tempt the consumer to resort to it in place of more dearly bought animal food. Another consideration should also induce us to produce the best cheese and butter at the cheapest possible price, namely, the exigencies of a foreign competition, importing into our markets goods of a high order at a low rate. We can only maintain our position as butter and cheese makers by producing an article which shall at least be equal, if not superior, to foreign produce, so as to enable us to demand a price that will repay the farmer. English make, if it only equals the American and other cheese, will always have the preference with English cheese eaters. But we have to deal with several difficulties in our competition. Rents are high, labour is scarce, and the cost of production is necessarily great. To make best cheese and butter of an uniform quality requires unremitting care and a commanding skill. The cows that produce milk must be kept in a proper manner and their food properly prepared. The aid of science must be invoked to help us in the proper management of milk when undergoing its conversion into butter and cheese, and last, and not least, by making use of the most fitting and efficient dairy appliances. With the latter object principally in view, we have been induced to write this paper.'[18]

The articles and reports devoted to dairying are not altogether without their lighter side, at least to a modern reader. It is pleasant to know, for instance, that 'In Class 1 (Old Bulls)... the first prize was awarded to Mr Treffry's "Sir Colin", a very neat and regular animal, with a mild and placid countenance'.[19] One also feels affection for 'Annette', and 'Lady Carew', who did well at the Bath Meeting in 1877. 'Mr Butt's "Annette",' it was reported, 'which took second honours, seemed a remarkably nice heifer, good and level at well-nigh all points; while Mr St John Ackers' "Second Lady Carew", which took the reserve, and subsequently received fourth prize at Liverpool, displayed nice hair and character, with grand middle and lovely bosom.'[20] By 1879 "Annette" had, understandably, developed. She was now 'cylindrical in form and massive'.

The Livestock Report for 1881 refers to 'that wonderful heavy-fleshed, massive and level young matron, "Lady 3rd"', and adds that 'Mr Carwardine's "Pretty Face" is almost perfection, and for scale, symmetry, wealth of flesh and beautiful appearance, she is not to be surpassed; truly a diamond of first water. The daughter of the prize bull, "Oxford"; her square, compact form affords a plum-like fulness, and a Sussex heifer with better loins has never been seen.'

Colonel Luttrell's reports on the horses were also very fine. He never minced words. In the Hacks Class at Bridgwater in 1883, for instance, we learn that 'Mr

Foster got the blue ribbon with a gaudy grey mare, which appeared to be helpless with its hind legs', while the 3-year olds were 'a moderate class, with the exception of the two first—Mr Yeo's "Amble Lad" and Mr Lang's colt—the former very light under the knee, and a sprawling goer; the latter well put together, with plenty of growth in him, and likely to tumble into a hunter, a position I doubt the winner will ever arrive at.'[21]

One is also pleased to note, in 1875, a new rule for the Poultry Section of the Show: 'All Eggs laid will be destroyed'.

The Society was certainly providing its Members with a rich variety of practical information and opinion, at the Show and in the *Journal*. The best way of deciding whether they liked what they were offered is to see if they continued, year after year, to pay their subscriptions. There are two ways of trying to discover this. One is the rough and ready method of comparing the membership totals in, say, 1863 and 1873. If there were more members in 1873 than in 1863, then presumably the Society was on the right road; if there had been a decline, policy needed careful thought. By this standard, the Society can be reckoned to have done pretty well. There is, however, another method; to see whether individuals who were members at one date are still members at another. This, it must be admitted, has drawbacks of its own. Even the best of Societies cannot prevent its members from dying or from passing into a senile retirement. But, even so, a study of membership by names, rather than by totals, can provide useful information.

In 1853, there were 44 Members whose names began with the letter B. 21 of these are still to be found in the 1863 list. In 1863 there were 126 names, and, of these, 55 survived into the 1873 list. This means that between 1853 and 1863 the Society held on to 47·7 per cent of its letter B members and between 1863 and 1873 to 43·6 per cent. For letter S, the percentages were 43·7 and 37·2. It is therefore not unreasonably to conclude that, since there was a high turnover on a larger membership, the Society had to work harder in order to keep the subscription income from declining, a feat which it usually accomplished, although there were, inevitably, some fluctuations from year to year.

When one goes beyond the broad totals of expenditure and investigates the details, the accounts yield a good deal of pleasure and not a few surprises. At the 1863 Annual Meeting and Show in Exeter, for instance, £33 9s 2d was paid to the police, £109 6s 0d to the band and £49 5s 0d to the South Kensington Museum. Why, one wonders, did the band cost so much and the police so little? Ten years later, however, everything had gone up. The police cost £103 13s 6d, and there were new and modern-sounding items, such as Rosettes, £7 11s 4d and Badges, £6 4s 6d. The Secretary was getting the respectable salary of £250 and the Editor £150 and, a sign of the times, there was 'Payments to Authors £88'.

The Society entered its second century with 1,033 members and £10,000 invested in Government stock, a very different situation from twenty-five years earlier. A merger with the Southern Counties Association had been arranged in 1868 and, after a hundred years of very fluctuating fortunes, the Bath & West was now unquestionably the strongest local society in Britain. At the Centenary Meeting in Bath, the Secretary, Josiah Goodwin,[22] said, 'The founders of the Society were among the first, if not the very first, to promote the welfare of the English people by a systematic co-operation between the tillers of the soil and the cultivators of literature, art and science. They recognised the intimate connection between agriculture, manufactures, and commerce. They took the trouble to inquire into facts in various parts of the British dominions; they recorded the experience of practical farmers; they endeavoured, according to the knowledge of those days, to dive into the principles of nature, and illustrate facts and experience by science.'[23]

The Patron, the Prince of Wales, was not present, although he had been at Guildford in 1871, when the Society was 'honoured for the first time by the presence of Royalty', and at the Mayor's Banquet, supplied 'on a scale of profuse liberality by Fortt and Son, of Milsom Street', the Mayor, Jerom Murch, proposed the health of the Patron in a masterly fashion. 'We can all imagine an event,' he said, 'which would have caused it to be drunk with greater enthusiasm. There was a time when some of us had hoped the Prince and Princess of Wales would have been induced to honour us with a visit. We knew that the engagements of his Royal Highness were numerous, but we thought the exceptional circumstances of the present occasion would have weighed with him. This is the Hundredth Anniversary of a great and very useful Society. The Prince has long been its Patron; as Duke of Cornwall he has considerable property in the neighbourhood of Bath; moreover, to what worthier objects could the Heir to the Throne give his attention? Yet, notwithstanding all this; notwithstanding the earnest wish of a large and loyal city; notwithstanding the kind efforts of the Council of the Society, headed by its President and the Lord-Lieutenant of the County—his Royal Highness has found himself, for reasons stated in a very courteous letter, unable to spare a few hours for the visit. Our feelings, however, must not go beyond disappointment; our loyalty is not for times of sunshine only; it can also be firm and strong, though perhaps not so exuberant in the shade: I give you, therefore, "The health of the Prince and Princess of Wales and the other members of the Royal Family".'

Sir Thomas Dyke Acland proposed 'the Memory of the Founders of the Bath & West Society', and reminded the guests that 'the ostensible founder of the Society was a poor Quaker, of humble birth, of the name of Edmund Rack', who 'asked the nobility and gentry of Somersetshire and the surrounding counties to assemble in the city of Bath and take active measures for the improvement of the condition of their fellow-creatures and the cultivation of the soil of the kingdom.

Their desire was not to increase personal influence or profits, or even to raise wages; but they wished to establish the principle that agriculture was intimately linked with the commerce of this country. They considered it necessary to tell the squires of the country that there was an indissoluble connection between the prosperity of the town and country.'

The Show was held near Beechen Cliff, 'but the deplorable calamity which occurred on the third day, in the fall of the Widcombe Bridge while thronged with persons on their way to the Showyard, threw a sad gloom on all after-transactions. The lamentable occurrence demands special reference in this place, from Mr John Azariah Smith, one of the principal exhibitors of Devon cattle, and an old member of this Society, being amongst the victims. Thankful are we to state that his life was spared, but it was only after suffering some weeks under special surgical treatment and with the loss of his leg.'[24]

Joseph Darby reported euphorically on the Implements Section of the Show. 'The growth of the Implement Department at our great Agricultural Shows,' he believed, 'can only be attributed to two causes—namely the increasing dependence of modern farming on labour-saving machinery, and the studious endeavours of our mechanics and agricultural engineers to keep alive the spirit of invention, and bring to perfection every device calculated to relieve husbandmen of toil. Only by reviewing their services in the past can we become duly sensible to the magnitude of the work in which they are engaged, and of what may be expected of them in the future. They scheme, try experiments, and ponder over many a wearisome task in secret, encountering numerous disappointments and failures; but a noble object gives them patience and fortitude, and makes them regardless of toil. This is to accomplish something truly serviceable in their day and generation, worthy of being held in remembrance by posterity and of securing fame and the gratitude of millions. Our engineers in the past half-century have, indeed, run a brilliant career, and the transformations effected by them in almost every department of agricultural labour are truly marvellous.'

'Viewing the entire field of agricultural labour,' there were, in Darby's opinion, only three items of machinery still missing, 'an efficient turnip-thinner, a good potato-raiser, and a perfect self-binding reaper'. These, he was sure, were on the way. 'My faith is so great in the power of our engineers to cope with difficulties, that I have not the slightest doubt we shall soon have machinery fully answering the purpose. They never give up a task to which they have once applied themselves, and are so arduous in achieving gigantic undertakings, that the words of the poet, with a very slight variation, seem almost more applicable to them than to any other class:

"Men the brothers, men the workers, ever planning something new,
That which they have done but earnest of the things they mean to do" '

This was Victorian England at its most confident. But a show which could now reckon to attract 80,000 and more people could no longer rely entirely on machinery and cattle to satisfy them. Experiments had to be made with different kinds of attraction, all of which found their critics to begin with. A horticultural section, for instance, would seem an innocent and natural enough innovation, but there were those purists who were against it, on the grounds that horticulture was not agriculture.

The battle over horticulture marked the first real struggle between two temperamentally different groups on the Council, the entertainers and the instructors, a rivalry and a lack of mutual understanding that has never completely disappeared and, one hopes, never will, since the people whose main wish is to teach and inform keep the entertainers from being absurdly trivial and those whose aim is to amuse prevent the instructors from being intolerably dull.

'A Society which appeals to all classes, and so many varieties of taste,' the Council believed, 'may claim to be excused if it aims to gratify as well as to instruct. People who laugh at the introduction of flowers at an agricultural meeting need only to be reminded that they add to the attractions without taking anything away; and if the sense of the ninety thousand persons who visited the Bristol Meeting could be collected on this point, it would probably find expression in the well-known formula, "Let those laugh that win". Indeed, throughout the entire week the horticultural tent was one of the great centres of interest, and considering the high character of the flower-shows periodically held at Clifton, it must be accepted as a very high compliment to the Department that the local press, so far from disparaging the show, were unanimous in proclaiming its complete success. The collection of cut roses was probably without exception the very finest ever witnessed in the provinces, and the splendid specimens of stove and greenhouse plants contributed by amateurs and nurserymen afforded very decisive proof that at Bristol, at all events, the introduction of horticulture was no mistake.'[25]

The horticultural battle took place at the end of Maule's Secretaryship. He retired from active duties in 1866, although he continued as Honorary Secretary for another seven years. Josiah Goodwin, who succeeded him, was better fitted to deal with the Society's links with the general public. Like another journalist, Thomas Plowman, who took over in the 1880s, he was a man to whom public relations were the breath of life. Under their influence, the Society was gently and tactfully broadened out from what it had been, for a hundred years, a gentlemen's club, and led into the hurly-burly of mass-appeal. Not all Members relished this, but, as Goodwin and Plowman showed, there were ways of making the public happy and of contributing to farming progress at the same time.

During the last quarter of the nineteenth century the number of agricultural workers in the South-West dropped sharply. For the five counties of Cornwall, Devon, Dorset, Wiltshire and Somerset, the Census showed 120,000 in 1871, 106,000 in 1881 and 91,000 in 1891—a fall of 24 per cent in twenty years. The cause was a steady decline in arable farming and a transfer of land and capital to dairying and, to a lesser extent, meat production.

Wages were low, especially in Dorset and Wiltshire, but bonuses and overtime raised the figure a good deal in all counties. In 1894, the average rates for five districts were:

District	Weekly Wages s. d.	Weekly Earnings s. d.	Excess of Earnings over Wages %
Truro (Cornwall)	14 0	16 3	16·1
Crediton (Devon)	13 6	15 8	16·0
Langport (Somerset)	11 0	12 6	13·6
Dorchester (Dorset) ⎫	10 0	⎧14 6	45·0
Pewsey (Wiltshire) ⎭		⎩14 9	47·6

Without the expanding trade in what was well-termed 'railway milk'—milk taken from the country areas to the cities, dairy farming would have been ruined. The price of cheese and butter was low, but it would certainly have been even lower if all the milk produced had been sold in this form.

But the growth of the liquid milk trade had another and more far-reaching effect. As Professor J. P. Sheldon put it, in 1895, 'So long as all dairy farmers made their milk into cheese and butter at home, the question of quality in the milk did not force itself into the ordinary bucolic mind, but it has been brought into prominence by the demands of the milk-trade, and by the creation of large establishments for the manufacture of cheese, butter, or condensed milk, as the case may be. There is now the probability of milk being, both in the trade and in the establishments alluded to, paid for on a basis of quality, to be determined by the percentage of solids. This, indeed, though generally in a somewhat crude and unscientific way at present, is also the tendency of the period, and there is reason not only to hope but to believe that it will in course of time become established as a sound and appreciable element in the commerce of the dairy. This accomplished, the sequel will be seen in the greater care taken in the construction of dairy herds, with the object of securing not only quantity but quality of milk.'

The '80s and '90s were also distinguished by the realisation that by comparison with certain other European countries, Britain was lagging considerably behind in the provision of both general agricultural education and of the human and financial resources for scientific research, without which further progress would be greatly handicapped.

The Age of Showmanship

Horticulture continued to add pleasure and a note of luxury to the Shows throughout the 1880s, partly, no doubt, as a counterweight to 'the dark cloud of depression which still overhangs both agriculture and commerce.'[1] At Exeter in 1879 there were 'orchids, palms and other choice specimens', and at Worcester the following year 'orchids and other exotics, tree ferns of rare growth and development, and a perfect collection of alpine and herbaceous plants, together with hardy begonias'. In 1881, at Tunbridge Wells, 'in consequence of the Society's visit this year to the Garden of England, prizes have been awarded by the Society for grapes and asparagus. Another interesting feature in this department is the corn grown in the open and under the influence of the electric light by Dr Siemens.'[2] And at Cardiff, in 1882, the 'extemporized rock work' proved popular. At the 1882 Show, incidentally, the Society used a contractor for the last time until after the 1939–45 War. The following year, at Bridgwater, it began the practice of putting up its own sheds, under the supervision of Robert Neville, of Butleigh Court, who was given the title of Hon. Steward of Works. At the same time, Henry Spackman, 'for many years official Superintendent of the Showyard and the Works connected therewith', retired and became Honorary Consulting Surveyor, with an ex-officio seat on the Council. Meanwhile, the number of Show Committees had multiplied. There were now ten of them, including Railway Arrangement, Disqualifying, Contracts and Refreshment.

The retirement of Josiah Goodwin from the Secretaryship in 1882, because of ill-health, led the Society to advertise for a successor. They had, as a result, 1,073 applications, of which they eventually examined the testimonials of 340, and from the 340 they chose Thomas Plowman, who had been Secretary of the Oxford Society for the past fifteen years.

Plowman was born in Oxford in 1844 when Acland was thirty-five, Voelcker twenty-two and Rack had been dead for more than half a century. 'My father,' he wrote, 'in addition to holding several public appointments, was a bookseller, publisher and journalist combined, as well as a writer upon many subjects.'[3] When he was seven, he was taken to the Great Exhibition, where he particularly admired the Koh-i-Noor diamond, 'the turbaned Turks and the pig-tailed Chinamen.'[4] His father was the secretary of the Exhibition for Oxfordshire, and the family treasured the bronze medal and large engraved diploma presented to him by the Prince Consort in recognition of his efforts.

After leaving school, he worked first as a 'handy lad' at the Bodleian and then as a librarian in Oxford City Library for what he described afterwards as 'between ten and eleven comfortable years'.[5] His father was the University correspondent of the *Morning Post*. When he died, Thomas got the job and added it to his existing duties as a librarian. In 1867, at the age of only twenty-three, he became Secretary of the Oxfordshire Agricultural Society, which his father had been, too, and ten years later became Editor of *Jackson's Oxford Journal* as well.

He enjoyed both parts of his life, the agriculture and the journalism, but he was ambitious, and when, in 1882, the Bath & West Society advertised for a successor to Josiah Goodwin as Secretary, Plowman put in for the job, acting as his own publicity agent in a way that today might be thought hardly decent, but which was normal enough at the time. His contacts with the Press were excellent, as indeed they should have been, and he used them to the full. Newspaper after newspaper published what amounted to testimonials, saying what a splendid candidate he was and how fortunate the Bath & West would be if it appointed him. The Society's Library contains all the press cuttings. The item in the *Agricultural Gazette* was typical. 'He has been distinguished by great powers of organisation, prompt and clear account-keeping, and great personal courtesy and readiness to afford information, which would admirably fit him for so important a post.'

Two weeks before the Selection Committee met,[6] Plowman sent out a printed circular to all the members of the Council of the Oxfordshire Society, giving the names and addresses of the Council Members of the Bath & West Society and saying that 'he hopes he may have the benefit of the influence of anyone favourable to his candidature who may be acquainted with any of the Council of this Society, whose names are subjoined'.

It was a masterly campaign. At the Oxfordshire A.G.M. a few days after the Bath Committee had met to interview candidates, Plowman said—and this, too, was fully reported in the Press—'he knew that if he was successful in obtaining the office it would be almost entirely due to the Society. He had ample evidence of that when he went down to Bath and appeared before the Committee. The efforts that had been made on his behalf were then specially alluded to, and one of the leading London agricultural papers, which was kind enough to comment very favourably on his candidature, particularly remarked on what it considered was a strong point in his favour—that he was so enthusiastically supported by the members of the Society he represented.' As a token of their esteem, the Oxford Society, at their 1883 meeting, gave him 'a beautifully illuminated framed address' and 'a purse of 120 guineas', and the staff of the *Oxford Mail* 'an oak inkstand, handsomely mounted in silver'.

The Selection Committee was unanimous in appointing Plowman to the post. The Society saved £28 5s 0d a year by very cunningly amalgamating the offices of Secretary and Accountant. Previously, by paying the Secretary £350 and the Accountant £128 5s 0d, the total salary bill was £478 5s 0d. Plowman came for £450, to do both jobs. Goodwin continued as Editor and received £200 a year for

30. Thomas Plowman, c. 1890.

his services. When he died in 1890, Plowman successfully applied for the Editorship as well.

Never one to do things by halves, he sent a printed notice to every member of the Council.

'My Lords and Gentlemen,—

The Editorship of the Society's Journal has become vacant by the death of Mr Josiah Goodwin, who, until ill-health intervened, held that office in conjunction with the Society's Secretaryship. As you may see fit to re-unite these offices, I beg to offer myself for the one at present unfilled.

I have held several appointments of a literary character, and have, by request, contributed papers which have appeared in the published transactions of Various Societies. I was also Editor of a County and agricultural newspaper for some years.

Three years ago you were good enough to testify, by formal Resolution, your opinion of my fitness for the Combined offices of Secretary and Editor to the Royal Agricultural Society of England. I could not desire a better justification for venturing to address you now than is contained in the terms of that Resolution.'

At the same time, F. J. Lloyd, 'Chemist to the British Dairy Farmers' Association', was appointed Associate Editor, to strengthen the *Journal* on the scientific side.

Plowman took a very active part in the life of Bath. He was a member of the Council for many years, graduating eventually to Alderman and Mayor. He was a magistrate, and a member of the Bath School Board. He was a hardworking and popular member of the Bath Literary and Philosophical Society. Among his many lectures to this Society were *The Aesthetes: the story of a Nineteenth Century Cult*, reprinted in the *Pall Mall Magazine*, *A Court Lady's Letter Bag: Some Eighteenth Century Sidelights*, and *Free and Independent Electors and Their Representatives: some Election Reminiscences from a Non-Party Standpoint*. He published several books, partly autobiographical and partly, thrifty man that he was, collections of his lectures and newspaper articles.

He was a man who always set great store on his efforts being properly publicised. On December 7, 1885, he read a paper to the Farmers' Club on *Agricultural Societies and their Uses*. It was very fully reported in the Press and the Society's archives include twenty-seven folio pages of cuttings relating to it, including one from the *Journal* of the National Agricultural Society of Victoria, New South Wales. In 1887, he addressed the Bath Literary and Philosophical Association on *The English Drama: the Conditions under which it has flourished or decayed*. To judge from surviving quotations, it was a somewhat rhetorical piece. 'Let us,' he said, 'be content to accept the Drama as a gift from an all-bountiful Providence, which affords to many, what human nature stands solely in need of, an intellectual recreation for brains often jaded by the wear and tear of life's work.' The Drama, he felt, 'has sown its wild oats and settled down to a fairly steady life'. For this, too, there are many pages of newspaper and periodical reports, all carefully captioned.

Bath Literary and Philosophical Association.

The Next Meeting will be held at the Institution on
FRIDAY, FEBRUARY 1st, when

THOS. F. PLOWMAN, Esq.,

Will LECTURE on the

"Free and Independent Electors" and their Representatives; some Election Reminiscences from a Non=Party Standpoint.

SYNOPSIS :—The electoral life of a constituency with a chequered history.—The "good old days" of electioneering.—The manners and customs of the "free and independent electors."--Types of Candidates: the writer who was no speaker; the framer of polished phrases who could talk long and tell nothing; the interrogatory Q.C.; the "Parliamentary War-horse"; the captivating vote-catcher. – The arts, wiles, and pitfalls of electioneering —A canvassing experience.—A petition and a commission and what they discovered.—Parting advice to all politicians.

The Chair will be taken at 8 p.m. by The MAYOR.

The Admission to Single Lectures is 1s.

Members are requested to enter the Institution from the Terrace Walk.

MOWBRAY A. GREEN, *Hon. Sec.*,
5, PRINCES BUILDINGS.

31. Bath Literary and Philosophical Society: announcement of lecture by Thomas F. Plowman.

From the Society's point of view, it was almost certainly Plowman the Showman who made the biggest contribution. He was an excellent organiser and Shows were dear to his heart, but by 1900 he had come to see quite clearly that the old concept of the Show as something primarily intended for country people would have to go. The number of people earning a living from the land was getting smaller each year, and, as a result, shows had to be attractive to townspeople if they were to survive. 'The turnstiles would never click fast enough to pay expenses,' he wrote, 'if only the rural patron passed through them'. The city-dweller had to be given attractions that competed with what he could get elsewhere, but how far ought one to go in meeting this demand? In this difficult transition period between the old and the new Englands, Plowman left no doubt as to where he stood. 'I hope,' he said, 'that our shows may be able to fulfil their mission without following the example of some of the Colonies, where a large portion of the Show-yard partakes of the nature of a country fair, and where acrobats, contortionists, performing animals, *et hoc genus omne*, are allowed to disport themselves. I would substitute for these plenty of good music, floricultural displays and exhibitions illustrative of arts and crafts and village industries.'[7] Whether this is what the new public actually wanted is another matter, and, in fact, Plowman gave them some of both, just to be on the safe side.

He accepted that what he called 'jumping competitions', were now well-established in public favour and was glad to see that such competitions had not attracted only, as he put it, 'circus horses, trained to negotiate a series of show-ring obstacles and good for nothing else'. The quality of horses entered had gone up, and so had their value. All the same, societies such as the Bath & West did not exist merely or primarily in order to feed the public with 'jumping competitions'. They were a means to an end, not an end in themselves. And he then laid down his great principle: 'Anything not strictly agricultural must not be allowed to absorb the chief energies of a Society, but must be regarded merely as a help to provide the sinews of war for the promotion of more essential work.'[8]

Plowman was a great inventor and provider of 'sinews of war'. He carried out his duties for thirty-seven years, during a period when the Society had to learn or be persuaded to adapt itself to a new set of values and priorities, to the existence, influence, and, occasionally, the subsidies of a Board of Agriculture, to great changes in the economic situation and the educational system of the country, and finally to the upheavals caused by the 1914–18 War.

One of the best tributes to him was paid, not in one of the many obituaries that appeared after his death, but ten years earlier, in the *Bristol Times*.[9] 'Why not Sir Thomas Plowman?' the paper asked. 'Given all credit to the workers who preceded Mr Plowman, will anybody say that his advent at Terrace Walk did not mean a new era for the Bath & West Society? The quality of the work he has done cannot be over-praised. We do not think that anything but his own modesty has left him "Mr". If there be another reason, it must be a Prime Minister's oversight.'

Plowman understood very well that the best way to get the general public interested in agricultural matters was to provide an opportunity to see farming operations in action. One of the Society's most successful developments of this kind was the Working Dairy. This had been pioneered in 1880 by the Royal Agricultural Society at their Carlisle Show and in the following year, at their Show in Tunbridge Wells, the Bath & West Society embarked on a similar but larger venture, using the most up-to-date equipment then on the market.

'A Working Dairy was this year one of the attractions of the Showyard,[10] and a short account of it and its teachings deserves a place in the "Journal". The arrangement was similar to that of the Royal Agricultural Society's at its Carlisle Show, with a few slight modifications. The Dairy was a shed 40 feet long by 15 feet wide, paved with squares, set with a good run to a gutter all along one side connected with a drain, a double canvas roof, and at one end the engine-house and office boarded off from the dairy proper. A line of shafting ran the length of the building, driven by a 6-horse-power portable engine lent by Messrs. Brown and May, which also supplied steam to one of Barford's boilers, giving an un-limited supply of hot water; cold water being laid on from the Yard supply.

'The Aylesbury Dairy Company kindly acceded to the request of the Society to work the Dairy and supply plant and men, the Society paying for milk and cream, and selling all butter, most of which, in ½-pound pats done up neatly in a piece of muslin in a basket, readily sold at 2s per lb to visitors, many of whom returned day after day for a supply, their general comment being that good butter was unknown in the neighbourhood. One side and end of the shed were open to the public, the opposite side being enclosed and fitted with seats, to which admittance was obtained by payment of one shilling on half-crown days and sixpence on shilling days, so that those who really wanted to learn and closely inspect the machinery and operations might be free from the mere sightseer.

. 'The machinery brought by Mr Allender, the managing director of the Aylesbury Dairy Company, consisted of eccentric and Midfeather churns driven by power, and an American swing churn worked by hand, all of which, as far as we could see, were equally efficient, the art of butter-making being in the churner and not the churn. A butter-worker for mixing various sorts of butter together, for working salt into butter, or working and washing it out; other sorts of butter-workers for pressing the moisture out during the process of making; and a Laval's separator, which was kept pretty constantly at work, extracting the cream from the milk, much to the astonishment of many who had never even heard of such a machine before.'

After the success of these demonstrations at the Annual Show, the Society developed a system of butter and cheese-making schools, held during the summer months. The farms were very carefully selected and pupils lived in during the course. In the first three years, 1888–90, the Schools visited eighteen places and operated for a total of 342 days. 348 students had come for ten days, 61 for a week, and 41 for shorter periods.

32. Clevedon Butter School, c. 1890.

An article by Plowman[11] explains the arrangements. 'Under the scheme, as originally formulated, the Society organised and controlled the School, and provided skilled teachers and the necessary dairy appliances, the district visited undertaking, on its part:

1. To provide, free of cost to the Society, suitable premises (with a sufficient supply of pure water) for the School, including a room for a Working Dairy (to be heated in Winter) not less than 30 ft by 20 ft in size, to be available for a fortnight, with the option of extending the term to three weeks, should the Society desire it.
2. To provide, free of cost to the Society, sufficient Milk for use in the Dairy, the Local Committee receiving the produce in return.
3. To defray the cost of local printing and advertising in connection with the School.
4. To guarantee not less than ten students for one entire course of instruction.
5. To secure the services of a Committee of Ladies to assist in obtaining suitable lodging accommodation for such of the female students as may require it, and, generally, to supervise arrangements in connection therewith.

The instruction fees and receipts for admission of spectators were taken by the Society, and the proceeds arising from the sale of the butter made in the School were paid over to the Local Committees. The Society, however, having been assisted by grants from Parliament has, during the past year, been able to render more pecuniary help to the Local Committees, and has handed to them half the receipts for fees and admissions. With this exception, the original conditions have been adhered to.

The initiative in obtaining the advantages of the School for a district is usually taken by the local agricultural association or farmers' club, which places itself in communication with the Society, and formally assumes such share of the responsibility as falls upon the locality visited. The Society is thus brought into pleasant contact with kindred bodies whose organisations are available for the promotion of objects which both have at heart.

FEES

A complete course of instruction in butter-making extends over ten days, but students can attend for shorter periods, if the School arrangements allow of it, though they are not eligible to take part in the School prize-competitions, or to receive the Society's certificate of efficiency. The instruction fees are:

	£	s.	d.
For entire course of 10 days	1	1	0
For one week's instruction	0	15	0
For one day's instruction	0	5	0

The equipment was lent by the manufacturers. 'The School has been visited on several occasions by representatives of Her Majesty's Privy Council, and (later on) of the Board of Agriculture', Plowman reported. These gentlemen 'expressed themselves most favourably, both as to the quality of the teaching and the general work of the School. The Introductory Report for the financial year 1889–90, presented to Parliament by the Board during last Session, says:

"The most important of the movable Dairy Schools and Lectures were those provided by the Bath & West of England, and Southern Counties' Society. The travelling Dairy School organised by this Society, with the assistance of Local Committees, visited eight different centres during the year. In all of these, under the guidance of well-qualified instructors, butter-making was taught to regularly formed, well-attended, and appreciative classes." '

He goes on: 'The Inspector who visited this School was much pleased with all he saw; the butter made at the end of the first week's instruction being worth 3d or 4d per lb more than that made at the beginning. He reported that the students appeared to take the greatest interest in their work, and that one student who had ten days previously been quite ignorant of the principles of butter-making had, the day before his visit, made butter the quality of which it would be difficult to improve. He also added that the butter made at the School was very much superior to that shown him as a good sample of the butter of the district.

'This official expression of opinion has been emphasised by grants from the Government of £100 in 1888, of £300 in 1889, and of £250 in 1890. Such recognitions have been a great encouragement to proceed, and have enabled the Society to engage more actively in the work than it otherwise could have done.'

There was an enthusiastic tribute to the not-overpaid teachers, Miss Davey and Miss Barron, and to Mr G. Gibbons, the Society's Dairy Steward, who undertook the general administration. 'The Society pays its teachers £1 1s 0d per week each and their travelling expenses, and also provides them with board and lodging, but no payment is made to them during vacations.'

The first Cheese School was established in 1889, at Palace Farm, Wells. 'With the sanction of the Ecclesiastical Commissioners, in whom the property is vested, arrangements were made with the tenant, Mr C. E. Wickham, for the entire use and control of his dairy, cheese-room, and cheese-making appliances. The Committee purchased the whole of the milk yielded by his cows during the time the School was open, at a uniform rate of 6½d per imperial gallon; the whey being placed at Mr Wickham's disposal free of cost. Mr Wickham also undertook at a fixed rate of charge to provide board and lodging at the Palace Farm for the teacher and a certain number of students, and to find suitable bed-rooms near the School for those who could not be accommodated in the house.

'The Committee were fortunate in being able to induce Mr H. Cannon, of

33. The Dairy School of the Bath & West Society, held on the premises of the Express Dairy Company, Hampstead. From *The Pictorial World*, 28 November 1889.

Milton Clevedon, Evercreech, to supervise the School, and his eldest daughter to act as the teacher. Mr Cannon enjoys a reputation second to none as a maker of Cheddar cheese, and he has gained the highest honours at the leading Agricultural Societies' Exhibitions. Miss Cannon had been actively engaged in her father's dairy, and had made the cheese which gained the Jubilee Champion Prize in the celebrated competition at the Frome Show in 1887. She had the training and experience which especially fitted her for the post of teacher, and the result has entirely justified the Committee's selection.'

The complete course lasted four weeks, for which the fee was eight guineas, including board and lodging. The number of students was limited to four. The success of the Wells School persuaded the Society to start another one, at Frome, also under the Cannons.[12]

The Cheese School at Vallis Farm, Frome, was the centre of careful scientific observation throughout the seasons of 1890–93, both by Miss Cannon and by F. J. Lloyd, a qualified chemist and the Assistant Editor of the *Journal*. The data are set out in great detail in the *Journal* in its volumes for 1891–94. They represent the most thorough investigation into cheese-making practice so far undertaken.

Silage-making attracted a good deal of attention during the 1880s and 1890s. The 1883–4 *Journal* has a long article about it. The principle to be followed, in the Society's opinion, was 'the greater the pressure, the better the silage'. Experiments were being carried on in various places, especially on noblemen's farms, e.g. Lord Walsingham's at Merton, Norfolk, Lord Tollemache's at Pecforton Castle, Cheshire, and Earl Fortescue's at Castle Hill, North Devon. Silage-making, however, continued to be a rich man's hobby. 'It is not every one,' admits the *Journal*, 'that can afford to run the risk of committing his crops for months to the operation of influences at work unseen in the silo—influences which may prove favourable or otherwise, as the result of causes at present very imperfectly understood. Where so much has yet to be learned, prudent men, not over-burdened with capital, will be content for the present to watch and wait.'

At the Annual Show at Maidstone in 1884, 'The Ensilage Section of the Implement Department attracted perhaps more attention than any other. It was the first time that any appliance connected with the storage of green fodder had been exhibited in the Society's Showyard, and was probably the most complete collection up to that time made in this country.'[13] One of the more interesting pieces of equipment for this purpose was Messrs. F. W. Reynolds' patent wooden silo. 'This resembles nothing so much as a huge tun set on end and covered with a low conical roof. It was here shown for the first time. It consists of wooden boards planed on the inside, and jointed to suit the radius of the required circle. These boards are erected vertically, like staves of a barrel, and are held together on the outside by iron bands, which are tightened up by lugs and screws, the bands being in sections. Some of the staves are supplied with staples, in which the bands rest before they have been tightened.'

The Society carried out silage experiments at Maidstone in connection with the Show, when specially built silos were filled. Among the materials used was

34. Silage-making demonstration, Maidstone, 1884.

hop-bine, 'which cattle ate readily', when the silos were opened in December. Reports of the practical work were accompanied, in the 1884–5 volume of the *Journal*, by an abridged version of one of Dr Voelcker's last pieces of work, his Royal Agricultural Society paper, 'On the Chemistry of Ensilage'.[14] The combination of Voelcker's scientific work and the practical experiments carried out by the Royal Agricultural and the Bath & West Societies brought silage-making much more within the reach of the ordinary farmer. Permanent silos, built of stone or brick, cost up to £500, but 'towards the end of the year 1884, it was found that in all probability there was no need whatever for the erection of expensive silos for the purpose of pickling the grass, and making it into silage. What is now known as the stack-system, was introduced, and at the Brighton Show last year, Messrs. F. W. Reynolds and Co. exhibited a stack, which was described and illustrated in the last volume of this *Journal*. Since then the Royal Agricultural Society has offered prizes for the best stacks of ensilage, the result being that two stacks, made on a system patented by Mr C. G. Johnson, of Oakwood, Croft, Darlington, received the prizes.'[15] Johnson's system applied pressure to the stack by means of a network of wires and a ratchet.

Six different kinds of silage were made at Bower Ashton during the summer of 1884. Six 'substantial water-tight wooden silos were constructed, each having a capacity of 250 cubic feet'. The silos were built in a barn. Each load of grass and each finished batch of silage was analysed by Voelcker. Voelcker's scientific report occupies thirty pages of the 1887 *Journal*.

35. Maidstone, 1884. Stewards and officials:
Back row, l–r: Jesse Ellis; R. A. H. Seymour; E. J. Sanders; E. W. Williams; T. F. Plowman; H. G. Moysey; Jonathan Gray; Col. H. A. F. Luttrell; Col. E. H. Llewellyn; Col. Troyte; R. H. Bush.
Front row, l–r: Capt. J. C. Best; Maj.-Gen. F. E. Drewe; C. T. C. Acland; R. Neville Grenville; Hon. and Rev. J. T. Boscawen.

In 1885 Plowman read a paper on 'Agricultural Societies and their Uses' to the Farmers' Club. The meeting coincided with the Smithfield Show, which he always made a point of attending. His paper, published in the *Journal* of the Farmers' Club, was reprinted very promptly in the Society's *Journal.* Plowman was never a man to neglect the opportunity of making his thoughts available to the widest possible audience. In this paper, he stressed the need for co-operation between societies. 'As it is,' he observed, 'each one goes its own way without any regard to what its neighbour is doing, unless that neighbour happens to trespass upon what the other considers its own particular district, whilst every now and then new societies are formed in neighbourhoods whose wants would be better met by a more liberal support of those associations already in existence. It rarely happens that there is any combined action on the part of societies for the attainment of any particular end, and they have no recognition from Government as authoritative exponents of agricultural opinion. In Scotland, on the other hand, an intimate and friendly connection has long subsisted between the Highland and Agricultural Society and the local societies. The former society exercises a parental supervision over these, directs their efforts into appropriate channels, assists in the arrangement of their exhibitions and, as occasions may arise, affords them a needless multiplicity of shows, marking its disapprobation in the case of local societies formed in districts already provided for, by with-holding from them its district show premiums. By such means, uniformity of object and action and a fuller utilisation of resources are secured, and an effective machinery for the prosecution of enquiries is always at hand.'[16]

The Scots, the Prussians, the Danes, the Americans, all had a well-integrated system of agricultural societies, and the societies received large Government grants to help them in their work. In England, things were different. Here, the State took so little interest 'that it has never made an attempt even to procure a list of them, much less to ascertain what they are doing. In contradistinction, take the last issued Annual Report of the Commissioners of Agriculture and Arts for the Province of Ontario and you will find it contains a report from every agricultural organization in the Province of its work during the previous year. Every electoral district has its leading society, with the township societies working in connection with it. The Government collects and prints all their reports, including their statements of income and expenditure, the statistics they have collected during the past year, and the essays to which they have awarded premiums. These are issued by the Government in a well-illustrated volume, which also contains a summary, in the form of a report, on the whole work and leading features of the year, by the Commissioner himself. A fund of reliable information is thus provided, which is of immense service to all who feel a practical interest in the agricultural condition and progress of the Province. The societies themselves are kept up to the mark by the stimulus of publicity and criticism, and their usefulness is brought prominently under notice. The publications issued by the Government Agricultural Department of the United States are of the utmost value. In England, if it were not for the agricultural press, whose space is necessarily limited, little or nothing would be known by the country generally of what societies were doing, except in the case of the few which issue journals of their proceedings.'

Plowman was asking, in effect, the same questions that had agitated the minds of the Society during the dismal period of the 1830s and 1840s: how does a provincial society do useful and necessary work without duplicating the programme of a national society? And how does it stay solvent in the process?

By the end of the nineteenth century, the answers to both these questions had become fairly plain. The Bath & West would continue to make sense by specialising on those branches of agriculture and horticulture which were of particular importance to the region, by digesting the results of nationally-based research and passing these on to Members, by crusading on behalf of both popular and unpopular causes, and by nursing promising local experiments until they were big enough to support themselves and to find sufficient finance and encouragement from other sources. It would maintain its funds in a state of health by charging an adequate subscription—few societies of any kind in fact do this—by continuously monitoring public taste and developing new ideas to gratify it, and by dropping activities immediately it was becoming evident that their day had gone.

On the whole, the Society has done all these things rather well. Not that it has always made money, even when it deserved to. In 1887 the Annual Meeting was at Dorchester. The Prince of Wales attended, the weather was reasonable, and there was good local support, but the population within easy reach of Dorchester was simply not large enough to make the Show a financial success. Dorchester was not Bristol or Plymouth, places where, to use one of the Secretary's apt phrases, any fool could make a profit. Yet 'the Bath & West of England Society has never failed to possess the courage of its principles. While visits to large centres of population are from time to time imperative to its existence, it has never hesitated to incur certain loss whenever it might be thought incumbent upon it to hold its Exhibition in parts of its districts which could be benefited by an expenditure of money elsewhere accumulated.'[17] Or, in other words, the profit that was made at Bristol was used to subsidise Dorchester.

Quite frequently, it was necessary to give the farming community a jolt, to remind it that old methods and old attitudes were a sure road to bankruptcy. This kind of shock treatment was administered more than once during the late '80s and early '90s, in an attempt to persuade farmers to produce for a real, not an imaginary market.

In 1888, the Society published an article called 'Modern Pig Breeding', by Professor James Long, of the Royal Agricultural College. In it, Professor Long lamented the rearing and showing of over-fat pigs. 'It is only necessary to visit

36. Bristol, 1886. Shorthorn bull.

the Smithfield Show,' he wrote, 'where valuable pieces of plate and numerous prizes are annually distributed among animals which are simply animated rolls of lard. Such pork is not only extremely wasteful but most unsaleable, returning to the producer a far lower price per stone than pork ought to do, misleading the public and still further inducing breeders to make crosses with the stock these pigs represent to their own ultimate loss. When it happens, if it ever does, that the carcasses of the prize animals are exhibited,[18] the public will then be able to discriminate and to understand that the prize pigs they are so commonly accustomed to applaud are utterly unfit for their tables, and inferior in every sense to the pork they are accustomed to buy in the suburbs of London at a matter of $7\frac{1}{2}$d per lb.'

The situation, he insisted, had become ridiculous. Fat pigs were given the prizes at shows, but the public was demanding leaner and leaner meat. 'The difficulty of changing the present system,' he realised, 'is enormous. It cannot be expected that any step in a new direction would be taken by an exhibitor who would practically sacrifice every prize for which he competed under present conditions. The Judges have not the power, even if they have the will, for whatever occurred under the judgment of men determined to strike out a new line at one show would be undone at every other, inasmuch as the pig exhibitors would decline to send their exhibits for judgment under men who decline to arbitrate according to recognised customs. What then is the alternative? There is no possible course but that which ought to be taken by the leading Agricultural Societies themselves, who if they had the will to set about the work in right earnest should specify the type of animal which alone they were willing to encourage, and engage Judges who were willing to carry out the wishes of each Society.'

Abroad, matters were arranged differently. 'Very much of our pork comes from Hamburg. During the National Exhibition held there some years ago, we were shown over the enormous Chicago-like pig-killing establishment of Koopmann, whose brand is so well known in the London market. Here a thousand head are slaughtered daily, the majority of the animals coming from the dairy districts of Holstein and sometimes from Schleswig; but will it be credited that the whole of these pigs are of one special type, never fat, always sizeable, and as London buyers will know, provided with a proper amount of lean? It is one of the misfortunes of the English producer that he is combated by the foreigner, who never ceases to study these requirements of his customers which he is content to ignore. The Pig Breeders' Association will sooner or later have to take this question in hand.'

Sir J. B. Lawes continued the campaign, in the same volume, with an article called, straightforwardly, 'The Pig of the Future'. In it he gave pig-producers some basic information which he felt had so far escaped their attention.

'Many years ago, when I was staying with the late Sir Henry Thompson in

Yorkshire, he told me that the taste of the manufacturing population in his district was changing very much. There was a time when the great demand was for very fat Cotswold mutton, but recently the Down mutton, with more lean and less fat, was in demand. The explanation he gave for this change was as follows: Formerly a fat chop was put into the frying-pan with potatoes, and both were fried together. Lately, however, the artizan population had become much better off, and took their fat in butter and more costly forms than mutton suet. It is to somewhat similar causes that the demand, not only for pork and bacon, but also for beef and mutton, in which fat and lean are more evenly distributed, has of late years arisen. It is not by any means the case that the demands of the population for fatty substances have in any way decreased, but simply that they can obtain their supplies in other and more palatable forms. The agricultural labourer in my time, if he eat any meat at all with his bread, it was the fattest bacon he could obtain; and he rarely consumed any other form of fat. But this is all changed now, as he prefers meat with less fat, and takes his supply of fat in more palatable forms.

'There can be no doubt that butterine [a mixture of butter and margarine] and the various forms of artificial butter have contributed largely to this altered state of things. The farmers in this country are not altogether blameless in having brought about this unfortunate state of affairs, as the quality of the butter supplied to our large towns was far from being as good as it ought to be, and the consumer appears to prefer a tasteless fat, which is supposed to be butter, but is not, to real butter which has more or less a rank smell or taste. The more recent advice, which the farmer has received to meet the altered condition of the pig industry is to convert his lard into butterine and so compete with the foreigner in his own trade.'

Two years later the Society's Consultant Chemist was saying exactly the same in connection with butter. He noted that Danish, Dutch and German exports of butter to Britain had grown enormously, because these countries supplied a uniform quality. 'The town consumer, tired of getting one day one kind of butter and another day another, and possessed with the natural idea that if one kind be good, all the others, so different from it, must be bad, clamours to be supplied with an article that does not vary. The butter-merchant, knowing he cannot depend upon a regular and adequate supply of a uniform article from home sources, turns to foreign ones, where he knows he can find butter manufactured upon one system, and that the best. In countries such as Denmark, for example, not only are there travelling Teachers, but also scientific men continuously engaged in investigation on dairy matters, and travelling Inspectors appointed, by whom the trade interests of the country are carefully watched. At the present time the Danish Government have, resident in our own country, a chemist for this express purpose; and their Agricultural Society has, in addition to its general

Consulting Chemist, another who is separately and specially devoted to the dairy industry. How different all this is from what we find in England! Again, if we turn to our agricultural journals and papers, we find that the prominent writers on the scientific aspects of dairying are Germans; whilst among our own countrymen there are but few names of note. Is it strange, therefore, that combination succeeds while our independent and isolated action fails?'[19]

'Beat the foreigner' was an excellent motto for British farmers to keep in mind during their round-the-year activities, but something else was required if visitors in their tens of thousands were to be attracted to the Annual Show. Thomas Plowman showed remarkable initiative in thinking up new popular features each year, related, wherever possible, to the place where the Show was being held. At Swansea in 1892, for instance, there was a stocking-knitting competition, with 92 entrants. Here visitors could see 'two or three girls attired in Welsh costume, including, of course, the far-famed sugar-loaf hat' busy knitting throughout the day, one group taking over from another every couple of hours. 'The knitting, as a rule, was very even and good,' said the report, 'and though it seemed a pity that more attention had not been paid, in many cases, to the shape, yet there were many very serviceable and excellent stockings among those shown.'[20]

The Swansea Show featured a great novelty, oil-engines, which provided the noise and busy movement the crowds loved. 'It is only within the last two or three years,' the Society noted, patting itself on the back for its ability to notice a good idea when it saw one, 'that the adaptation of petroleum for the purposes of small

37. Taunton, 1895. Stewards and officials. l–r: Chick (waiter); G. B. Mildred: Marquis of Bath; T. F. Plowman; C. L. F. Edwards; H. B. Napier; Perkins (Valet).

motors has been put to practical test, but within that period the new idea has made very rapid strides. It is evident that an oil-engine has special recommendations as a motor for agricultural purposes, inasmuch as it has all the convenience of a gas-engine without the difficulty of obtaining a supply of gas. In rural districts this difficulty is one which frequently becomes quite insuperable, and the farmer who wishes, whether for a separator, or for any other machine requiring a comparatively low motive power, to employ a small motor, had practically no choice between a steam-engine and a horse-gear. If, as now seems established, the invention and rapid development of petroleum engines answers its purpose, there is no doubt that it will supply a distinct want.'[21]

At Bridgwater in 1883, there was an impressive display of brick and tile-making, the main local industries. 'Among the Machinery in Motion, Messrs. James Culverwell and Co. exhibited their well-known brick and tile-making machines in operation. One was a double-ended steam-power self-acting machine, capable of turning out from 12,000 to 14,000 bricks per day. This is very powerful, and is fitted with self-acting belt-gear which reverses the motion for alternate making, one plunger making pipes and bricks while the box last discharged is being refilled. A single-ended machine exhibited was of similar construction, but worked with one plunger, and making only 7,000 bricks per day. Shown at this stand was a new combined double-pug and roller-mill and brick-

38. St. Albans, 1896. Cider competition. The central figure is F. G. Fawcett.

39. Bristol, 1886. Stallion, 'Mormon'.

making attachment, invented by Messrs. Culverwell and Co. With this machine the clay is placed in one end, first pugged, then crushed between heavy rollers, again pugged, and expressed for pipe or other machine. It is driven by one belt from a portable or other machine.'[22]

At the same Show, there was an exhibit devoted to that important domestic scouring item, Bath brick, well-deserved, because, as was pointed out at the time, 'the town of Bridgwater has to endure the annoyance of turning out an article of wide use and extended fame, while another place gets all the credit. It is not known to many except Westcountrymen, that Bath bricks are produced at Bridgwater. These humble adjuncts of every well-ordered scullery are made from the mud of the river Parrett, the sea itself mixing the ingredients of the composition. A singular fact is, that only within a mile each way of the town is the mud of the right consistency for the purpose; a mile above it contains too much clay, and a mile below too much sand.'

This particular Show broke with tradition by having no Pigeon Class. 'Year by year,' it was noted, 'the Pigeon Classes have diminished, till the entry money must have been far from paying the prizes, and these, from the paucity of birds, occasionally went to undeserving specimens. We do not think that true fanciers, as distinguished from professional exhibitors, will regret the change. All pigeons should, at this time of the year, be kept at home attending to their parental duties.'

In the 1890s, great improvements were made to the amenities at the Show. An

40. Guildford, 1894. Display of applied art by R. Pratt and Sons, cabinet-makers and upholsterers.

impressive new entrance building, designed by H. B. Napier, one of the Stewards of Works, was provided. It was a permanent structure, easily moved each year from site to site, and it sheltered the turnstile-keepers and ticket-takers. 'It is, said the Annual Report, 'in the style of an Old English timbered building, with lattice windows and a gabled roof covered with antique tiles. A clock-tower rises in the centre, and the whole forms a distinctive entrance to the Showyard.'[23]

In an enclosure near the entrance there was now 'a neat row of sleeping apartments for the accommodation of those of the Stewards whose devotion to duty induced them to remain at night within call of the various departments for which they were responsible.' Postal and telegraph facilities were ready at hand 'and telephonic communication was also established with Bristol, Cardiff and many other places by the Western Counties' and South Wales Telephone Company, Limited, whose office was near the entrance. The protection of the Buildings in the Yard from fire was entrusted to Messrs. Merryweather & Sons, of Greenwich Road, London, who had a powerful steam fire-engine, a manual engine, and all other requisite appliances ready for use at any moment, which, however, were happily not required.' A detachment of the St John Ambulance Brigade was stationed ready for action in a marquee. This was the full-blown modern Show, with only the car-parking to be added at a later date.

Before the century ended, the Society produced details of one of the most am-

41. Southampton, 1897. Stewards' Pavilion. *l–r*: C. L. F. Edwards; H. B. Napier; G. B. Mildred; A. O. Sillifant; T. Latham.

bitious and forward-looking schemes in its history so far, the Bath & West and Southern Counties Dairy Training College. The plan for this was worked out by a special committee. The College was to have two aims:

'1. To provide instruction of the highest class in:
   (a) the practical details of Cheese and Butter Making
   (b) the science of Dairying in all its branches
   Under Section (a) it is intended to give Practical Instruction in Cheese and Butter Making, as at present carried on in the Society's Schools.
   Section (b) is more especially designed for Students who are desirous to qualify themselves as Teachers of the theory, as well as the practice of Dairying.
2. To carry on systematic investigations with reference to Milk and its products; and to conduct experiments with regard to the feeding of Milch Cattle, and Dairy Husbandry generally.'[24]

It might, it was felt, be better to have two Colleges, rather than one, to minimise travelling difficulties and to foster local allegiance. If this could be arranged, there should be one College in the South-Western group of counties,[25] and another in the Southern group.[26] The key to the whole plan was, of course, money and it was recognised that this would have to come from the County Councils.

The proposal was therefore 'that a County Council contributing financial support to the scheme shall have the privilege of nominating a certain number of

42. C. F. L. Edwards.

students (in proportion to the amount of the grant) at a reduced rate of fees. The situation of the College would be decided hereafter by a Joint Committee representing the Councils joining in the scheme and the Society. It is expected that the Council of the County in which the College would be situated would, in consideration of the special advantages accruing therefrom, make an additional and substantial contribution to its funds. County Councils joining the scheme would, of course, be represented on the governing body.'

A number of collegiate centres for agricultural instruction[27] had already been set up with Treasury grants, the first of which was made available in 1888. These grants were supplemented from 1890 onwards by funds placed at the disposal of County Councils by the Local Taxation Act and Technical Instruction Acts of 1890. Local subscriptions had also been forthcoming. The Bath & West Society was less successful in its attempts to get such a college established. The required degree of co-operation between local authorities was not obtained and the South-West did not, in fact, achieve its own Dairy College until after the First World War, at Cannington, near Bridgwater. One reason for the failure was the negotiations, known both to the Society and to the County Council, which the British Dairy Farmers' Association—of whom Sir Thomas Dyke Acland was Vice-President—were carrying on with the University Extension College at Reading, with a view to attaching its own Dairy Institute to the College. Efforts were made to discover some way in which the College might co-operate with the Society, but the problems proved too difficult and in the end the Society decided to continue with their existing Dairy Schools and to keep the County Councils informed of the details.

A further experiment of the same kind was the Travelling Farriers School, which had been established in 1895. The prospectus for this announced 'The instructor is Mr W. B. Blackall, master smith, late of Coleshill, Highworth, Wilts, who, previous to his appointment to the post, had won twenty-one prizes and several high commendations at shoeing competitions held by the Bath & West and Southern Counties Society and other Societies.

'The instruction is restricted to those who are already in the trade. This is essential, not only to avoid jealousy and ill-will, but because the instruction can only be given effectively, in the necessarily limited number of lessons, to those who have already acquired a fair knowledge of ordinary shoeing. The aim is to improve old hands rather than to teach beginners.

'A course of instruction, the fee for which is 2s 6d, consists of ten lessons. These are given at six o'clock in the evening as the pupils, having their ordinary work in the daytime, cannot conveniently attend before that hour. A class consists of four pupils, and, as the same pupils cannot always attend night after night, it is generally arranged to have two different classes, which are taken on alternate nights. The pupils are shown the correct method of shoeing every kind

43. Taunton, 1895. Bath & West Shoeing Van. W. B. Blackall,
the instructor, is the bearded figure in the back row.

44. The Prince of Wales at the Dorchester Meeting, 1887.
From the *Illustrated London News*, 11 June 1887.

of horse they are likely to have to deal with, and how to adapt shoes to abnormal conditions of feet. A typical collection of shoes and hoofs is always on exhibition at the School, and the explanations given of them are much appreciated.

'Forges, iron, and all the necessary tools and appliances are provided by the Society, and are contained in a van, which is moved about from place to place, so that the School may be brought within easy reach of the smiths of any particular locality in the country.'[28]

The plan for the Dairy College was a logical expression of the Society's aims, which had been re-defined in 1886 in the report of a special committee, under the chairmanship of J. E. Knollys. This report recommended that the Society should extend its activities in four directions:

'(a) The establishment of a system of practical experiments to test the advantages, or otherwise, of the use of artificial manures, on corn and grass, on land in ordinary farming condition, based on the results obtained from the Rothamstead and Woburn experiments.

(b) The examination and testing of any new process for dealing with agricultural produce.

(c) The improvement of dairying.

(d) The collection and publication of information on new systems of cultivation, routine of crops, or other efforts which are being made for the profitable cultivation of land under probable low prices of corn.'[29]

These guide-lines continued to control the Society's activities until the outbreak of war in 1914 introduced quite different conditions under which farmers had to operate and brought Government into agriculture to an unprecedented extent.

This chapter is concerned with the period between 1900 and the end of the First World War, two decades during which the Government and the political parties found themselves increasingly aware of the necessity to take measures to improve the lot of working-class people and to make the division between the Two Englands of rich and poor a little less sharply defined and less brutal.

It was a period of pioneering social legislation—the Unemployed Workmen's Act of 1905, the Trades Disputes Act and the Workmen's Compensation Act of 1906, the introduction of Old Age Pensions in 1908, the Labour Exchanges Act of 1909, the National Insurance Act of 1911, Fisher's Education Act of 1918. It was also a period in which the Labour Party was established, the Transport Workers' Federation and the National Union of Railwaymen were formed, the Russian Revolution took place and the national boundaries and political systems of Europe were transformed by the outcome of the War.

The war itself changed the thinking of the agricultural community in many ways. Farm labourers who were conscripted and sent abroad to share the common miseries of men in their age group developed new attitudes and new aims as a result of living at close quarters with former industrial workers. Their traditional passiveness and acceptance of the rural social order could not survive the war. Many of them never went back to their old jobs and those who did saw their work and their employers in a different way.

This undermining, if not yet withering away, of a pattern of thinking and living which had for generations seemed normal and almost Divinely ordained was to some extent foreseen and understood by the curiously and misleadingly named Corn Production Act of 1917. The Act, which covered the whole situation and prospects of British agriculture, was in several parts. The first dealt with prices and the second with wages. There was provision in the Act for a Wages Board, to fix minimum rates.

An assessment of the Act by the Hon. F. D. Acland, published in the 1917–18 volume of the Journal, *showed what a remarkable amount of agreement there was between the parties on both its assumptions and its provisions.*

'All who had before the war any real knowledge of rural conditions,' he wrote, 'were agreed that the low rate of wages in great parts of England was not only keeping agriculture in a low state of development, but together with bad housing was smothering vitality and energy out of village life. Though for ten years before the war agriculture had been becoming more prosperous, little of the extra prosperity had been going the labourer's way, as the rising cost of living pretty well balanced increased cash wages. Emigration of the best men from the coun-

try went on steadily, and a vicious circle was established in that poor wages and conditions produced a type of worker which sometimes was not worth better wages, and sometimes could hardly have taken advantage of better conditions had they been established. This condition of things must be broken through if our home country-side is to be a strength to the Empire. Agricultural labourers who have served with the Forces will have learnt that it is not so difficult to move from one place to another as it may have seemed before the war, and only the certainty of good wages will bring them back. And we need a large new agricultural population in addition to what we had before, and only a certainty of good wages and conditions will attract it. This, at any rate, was the general feeling. So there was no dissentient voice in the House of Commons as to the necessity of minimum wage legislation.'

Wages, however, were not everything. After the war, at least the younger farm workers wanted a kind of life for themselves and their families which bore some resemblance to what was available to people who lived in towns. And this no Act of Parliament could give them.

Foreign competition and war as incentives to efficiency

In the *Journal* for 1906–7 the Society took the curious step of publishing a long report by J. H. Towsey, the British Consul in Milan. There is no evidence that Mr Towsey had any West Country connections, and his article was about the sins of British manufacturers, rather than of British farmers. The Society presumably decided to print it on the grounds that no opportunity should be lost of driving home the message that we were being outsmarted by the foreigners, and that it was high time we did something about it.

The occasion of the article was topical enough. An international exhibition was held in Milan in 1906, to celebrate the completion of the Simplon Tunnel. There was an Agricultural Pavilion, with Germany and France prominent among the foreign exhibitors. Eight British firms were showing implements. All received awards, but, in Mr Towsey's opinion, this was a pitiful effort. 'I may remark,' he wrote, 'that to one accustomed to the comprehensive displays at home, the British representation of agricultural implements at Milan was painfully incomplete. English ploughs, drills, harvesters, chaff cutters, etc., were conspicuous by their absence, and the question was naturally forced upon one's attention as to why there should be no trade in these implements between Great Britain and Italy.

'It is apparently the old story, constantly reiterated in the reports from British Consuls abroad: failure on the part of British manufacturers to adapt their goods to the requirements of foreign customers, and want of attention to important matters of detail in the packing, invoicing, and delivery of consignments.'

There was, he reported, a great demand in Italy for ploughs and drills, but 'at present this trade is entirely in the hands of the Germans, who have studied and are continually studying the local requirements. English ploughs, so far as Italy is concerned, are too well made and cost too much. They are too heavy and are not of the shape required for Italian agriculture.'

All this was no doubt true. The English have always tended to regard the habits and tastes of foreigners with scorn and suspicion, and to believe that to agree to sell them British goods is to confer a take-it-or-leave-it favour upon them. There has been an equally John Bull-like reluctance to believe that foreign imports to Britain ever sold on their merits. If the price was lower, it could only be because they had the benefit of a heavy subsidy. If the British public bought them at all, it was the result of cunning misrepresentation and it was only a matter of time before the foreigner's duplicity was discovered and his customers came to their senses.

A perceptive article[1] by A. T. Matthews, published in the 1907–8 *Journal*, illustrates this curious British failing, as it affected sheep farmers. Matthews observed that between 1886 and 1906 our imports of mutton, mostly from Australia and New Zealand, had increased from 635,447 to 4,082,756 hundredweights, and that the 1906 figure represented nearly $5\frac{1}{2}$ million sheep. All this foreign mutton depressed British sheep-farmers, and Matthews had a story to show the kind of personal decision which could result. A young farmer of his acquaintance 'was paying a visit to London, and, amongst other sights, inspected the docks, where a large cargo of New Zealand mutton had just arrived. He was shown over the ship which had on board, I believe, some 200,000 carcasses. He had before read of such things, but the actual sight created such a panic in his mind as to engender the utmost despair of the future of sheep farming in England. He was the occupier of a capital farm, cheaply rented under a good landlord, and, moreover, was the possessor of a very nice flock of pure bred Oxford Downs, good enough for ram breeding. Feeling convinced, from what he had seen, that the knell of the home sheep breeder had sounded, he went home, gave notice to leave his farm, sold off his stock and emigrated to British Columbia, a step which he has ever since bitterly regretted.'

What this reluctant emigrant had overlooked was the one fact which might have kept him at home, that during the first five years of the new century, the average price of best mutton on the London[2] market was higher than it had been forty years earlier. Imports had not lowered the price of home-produced mutton, which was regarded as a superior article. 'All the imported mutton,' Matthews pointed out, 'has simply gone in extra consumption. Being considerably cheaper than our native produce, our working classes appear to have come forward as consumers, and the only effect of foreign competition in mutton has been to create an entirely new class of buyers to whom mutton was before a prohibited luxury.'

Matthews' argument was probably sound, so far as mutton was concerned, but bacon was another story. An article by John M. Harris,[3] 'The Shortage of Bacon: Cause and Effect,[4] drew attention to a real shortage, caused, he believed, by the dumping of American and Danish bacon, which lowered English and Irish prices, since at that time it was possible for home-producers to meet nearly the whole British demand.

In the opinion of one well-informed observer,[5] the British cheese-making industry showed a deplorable lack of initiative. 'The demand for our finest English makes,' he declared, 'is much greater than the supply,' and for this reason it was France, Italy, Holland and Switzerland, not Britain which supplied 'the whole of the fancy varieties which find such a prominent place on the tables of the wealthy people, as well as of hotels and restaurants in all parts of the country'. The situation was not improving. It was, in fact, getting steadily worse. 'Twenty-five years ago one of the most famous varieties, Leicester, was superb—excelled by none, equalled by few others. To-day it does not exist, modern Leicester being of an entirely inferior character, nor have I seen a single specimen of the old type at Islington for a long series of years. The character of Gloucester and Derby cheese, neither of which are of high rank, does not enable either to take the place upon the London market to which they are presumably entitled, nor to obtain regular quotations in the weekly returns. Farmers, nevertheless, produce them, although at many shillings per cwt less than Cheshire, Cheddar and Stilton, and lose money in consequence. A large proportion of Stilton cheese is very inferior, and although the best quality realises 1s to 1s 1d per pound wholesale, large quantities sell at 6d to 7d. Wensleydale, although like Stilton, it is infinitely superior to Gorgonzola, is entirely superseded by the latter, which has so ready a sale in this country, and is frequently found at the Dairy Show and other agricultural exhibitions.'

There was no reason at all, he believed 'why fine British cheese of all the leading varieties should not furnish an export trade, when our own demands have been met, as easily as Italian Gorgonzola and Parmesan, or the Gruyère of France and Switzerland.' But such a development would demand a completely new concern for quality and a determination to go for the top end of the market, attitudes which James Long found seriously lacking among British producers.

Until the beginning of the twentieth century, a nation's standard of living had been assessed very largely by the annual consumption of meat and, to a lesser extent, butter, per head of the population, but by the outbreak of the First World War other types of foodstuff were being taken into the reckoning. One of them was sugar, and in an article written for the *Journal* in 1910 Charles Kains-Jackson went so far as to claim that sugar consumption could be used as an index of national prosperity. He provided a table, showing that the Australians ate or drank 129 pounds a year each, the Americans 84 and the British 81. The other countries dropped away down to Italy, with only 7. The figure for India was 44 pounds and for Germany 36.

Kains-Jackson's article was written at the beginning of the British sugar-beet industry, when the main problem was to persuade sufficient farmers to grow beet, so that a factory could keep going. The market possibilities, he assured his readers, were excellent and they could devote acreage to the crop with every confidence of getting a good income from it. 'The United Kingdom reaching the standard of Australian prosperity may, it seems, increase its sugar consumption by 48 lb per head; Canada, coming up to the United States, may increase its head of sugar by 9 lb and so on. The East India Association has clear justification for its hope of a vast trade in sugar in India. National leanings have a little to do with the use of sugar, but probably not much. Women and children, a great majority of the population in every land, like sugar, and want of means is probably the main cause of Italy showing so low a figure. Prosperous little Switzerland, receiving the tribute of all who love its Alps, heads the list of European consumers outside the United Kingdom, yet its people being a Franco-Germano-Italian blend, its theoretical figure should be 25 lb, not 65 lb. Not race, but purse, may fairly be held to control the yearly use of sugar in any given land. The whole subject is almost purely economic; for present purposes we have to assume no more than that in the ideal, well-fed, well-nourished state, the yearly consumption of sugar would at least reach the Australian standard of 129 lb. The Australians as a people are not adipose, they are not indolent, or in any way the victims of saccharism. They find in sugar, driving power at a moderate cost, and other countries are likely in this respect to follow their lead.'

In general, the papers published in the *Journal* maintained a very high standard during the first two decades of the present century. Some of the technical papers especially were outstandingly good,[6] but a not inconsiderable proportion of them were reprinted from other publications. In the 1912–13 volume, 84 pages were devoted to reprinted articles, which seems excessive, however high the quality of these articles may have been.

Eccentric and perverse articles were very few, and when they do occur they usually have sufficient charm to justify their inclusion on that ground alone. In 1910, for example, there was a contribution[7] from Professor J. Wrightson, advocating the use of cows for draught purposes. This, he said, could be done 'without detriment to their milking properties,' but to convince British farmers of the possibilities 'it might be possible to import trained cows in order to demonstrate to thousands of spectators the strength and docility of the animals'. Much, he thought, 'might be done by a Society for promoting Cow Labour, but until one is formed the Board of Agriculture and Fisheries might do something to assist the idea, or at least, to test it'.

This was a project which came to nothing, but no editor is sufficient of a genius to be able to publish only winning ideas. Plowman, however, backed remarkably few losers. During the early years of the present century, the *Journal* was quick to notice important new trends as they began to emerge. The 1905 *Journal* contains a good example of this, an article by W. J. Hosken on artificial incubation, illustrated by excellent photographs taken by W. M. Martin, of Redruth, which showed embryos and chicks at various stages of development.

'I anticipate,' said Mr Hosken, 'that in the next few years there will be a considerable advance in the number of those who use incubators,' a forecast which can almost be described as the agricultural understatement of the century. At the time he was writing, however, poultry-keeping in Britain was still carried on, for the most part, on a very small scale, often as a hobby for the farmer's wife. Weighing up the evidence and the probabilities, Hosken nevertheless concluded that 'modern methods had come to stay'. The increased interest being shown by farmers in poultry-rearing was, he believed, due mainly to publicity, 'to the literature that has been published on the subject, to the encouragement given to it by various County Councils, and more particularly to the importance attached to it by the Board of Agriculture'.

Technical changes were inevitable. Artificial incubation, in place of broody hens, was one of them and housing was another. 'The stone-built poultry house, hitherto part of the equipment of the farm, will not figure so conspicuously in the future, being superseded by the portable wood fowl house of recent design.'

The Society also drew attention, by means of carefully selected articles in the *Journal*, to actual and desirable improvements in dairying. It is difficult for a modern reader to realise how low the standards of milk-production were in the years just before the First World War. The notorious 'town dairies' were still in existence, wretched shacks in which cows spent the whole of the year, never exercising, never grazing and fed on very dubious material. 'In many of these places,' A. T. Matthews reported in 1912, 'the fresh air is jealously excluded, under the impression that the warmer a cow is kept the more milk she will give, and it is quite obvious that they must be hot-beds of disease. In these days of easy transit for milk there is no necessity whatever for the existence of such dangerous things as cowsheds in towns, and the sooner they are abolished the better it will be for the milk-consuming population, for the dairy farmer who carries on business under natural conditions, and for the increase of our stock.'[8] Dairy farmers were increasingly prosperous, not because they had become more efficient, but because of a steady rise in consumption of milk. The quality of most milk remained poor, however, and there was no legislation to force an improvement. 'A large proportion of the milk sold is as poor in its fat contents as it used to be,' reported James Long in 1914,[9] 'and there are spasmodic efforts so as to alter the law that low quality may be maintained. Again, there has been no great improvement in the methods adopted for the maintenance of the cleanliness of milk, either by the employment of more suitable pails, by the use of the milking machine, or by milking in the open air. Milk is still badly cooled, and large quantities are spoiled in the hot summer months, while the imperfect management of pastures and the passive rejection of forage crops, which play so great a part in the economy of the farm in other countries, are continual matters of regret. The better price of milk is owing to the increasing demand of an ever-growing population, and to the larger employment of milk as a food. The number of cows, too, has failed to keep pace with this growth; for the milking cattle of 1913 were but few more than they were in 1891. The result has been that large numbers of farmers, in order to sell milk, have abandoned the manufacture of butter and cheese, and so our imports of these materials are continually increasing. If, however, there has been no marked improvement in the quality of milk, there has been a distinct gain in the quantity yielded by cows, and it is probable that the average throughout the country, which was estimated at 430 gallons twenty years ago, is, to-day, from forty to fifty gallons more.'

Reading articles such as this, one is often tempted to feel that institutions such as the Bath & West Society were doing no more than preach to the converted, and that the majority of farmers remained stubbornly fixed in their backward ways, picking up perhaps one new idea every ten years. This is to be more pessimistic and cynical than the facts deserve. Change in agriculture has never been exactly breath-taking, but the average standard of farming practice in, say, Somerset or Dorset, was certainly higher in 1910 than in 1860. Farmers are influenced by their neighbours, by what they see at shows and demonstrations, and by what they pick up from journals and magazines. The innovators will always regard the routineers as absurdly conservative, and by the time their ideas have become common practice they are all too likely to be dead or to have moved on to something that excites them a great deal more.

Since, in any organisation, the conservatives will always be in a large majority, success depends on having a small core of people who are both energetic and diplomatic, who understand the value and importance of new ideas without overrating the eagerness of their fellow members to abandon old habits. The Bath & West Society has been extremely fortunate in possessing more than its fair share of people who were genuinely interested in progress, but sufficiently well-placed socially to be trusted and respected. A number of these key figures died between 1900 and 1920, and many who might have been their natural successors were killed during the 1914–18 War, so that the 1920s and 1930s were, perhaps inevitably, a rather fallow and unexciting period in the Society's history. This was particularly unfortunate, since the years between the Wars were characterised by stagnation and depression in British farming, and the Society was not well placed to provide the frequent injections of vigour and optimism the situation demanded.

Obituaries of prominent and useful people are to be found in every volume of the *Journal* in the immediate pre-war years.[10] One much missed link with the past was Capt. the Hon. John Charles Best, RN, who died in 1907. He joined the Society in 1868 and succeeded his father, the Hon. and Rev. Canon Best on the Council in 1871. He was elected a Vice-President in 1898. In 1912 the Secretary had quite a catalogue of such losses to report. There was Col. Wyatt-Edgell, 'who for long, as a member of Council, and especially as a Steward of Arts, rendered the Society valuable service'; the Earl of Onslow, 'for many years a Vice-President, and who filled the office of President in 1894 with much ability, and who also rendered yeoman service to Agriculture generally as a former Minister of that Department of the State'; and Sir James C. Inglis, 'who was always ready to help the Society to the best of his ability in connection with railway matters.'

A particularly sad loss, although he was not a member of the Society's inner cabinet or corps diplomatique, was George Lippiatt Gibbons, a notable figure in the public life of Somerset and one of the pioneers of the Working Dairy in 1885, and of the later Migratory Dairy Schools. He was, wrote Plowman,[11] 'a great believer in Sulphate of Ammonia, not only as a fertiliser, but especially as a personal pick-me-up. He carried a supply of it loose in his waistcoat pocket, and, when feeling tired and in want of a reviver, he took a pinch as though it were snuff, and always declared it did him "a power of good".' He supervised the practical work of the Cheese Schools: 'At the Bath & West Show at St Albans in 1896, the Princess of Wales (now Queen Alexandra), who was being shown round the Working Dairy by Mr Gibbons, observed with a smile, "the best butter, you know, comes from Denmark". "Pardon me, your Royal Highness," said the Steward, with a twinkle in his eye, "we think the best butter is made in England, but we have to go to Denmark for the best Princesses". The Prince of Wales (afterwards Edward VII), overhearing this, joined in with "ah, now you're buttering her".'

In 1911 James Rossiter died. He had been the Superintendent of Works at the Annual Show for more than twenty years. The value the Society placed on his services was indicated by his salary rises, £220 in 1901, £240 in 1902, and £270 in 1903. His place was taken by H. C. Ayre, who held the post until 1937 and stayed on until 1948 in an advisory capacity.

45. Plymouth, 1902: the Members' Pavilion.

Ayre's memories[12] are almost a history of the Society between 1910, when he began his probationary year as Superintendent, and 1960. He was born at Cleethorpes, and at the age of 15 he was apprenticed to a timber importer in Grimsby. Wanting to broaden his experience, he applied for a job in Bristol at the end of his apprenticeship. His employer at Grimsby did everything possible to persuade him to stay and, long afterwards, Ayre discovered that the plan had been to marry him off to one of the timber merchant's three daughters—there was no son—and to bequeath him the business.

In Bristol it so happened that he lived next to James Rossiter, Superintendent of Works for the Bath & West Show. The 1909 Show was in Exeter and Mrs Rossiter mentioned to Ayre that her husband wanted help in preparing the stands and fittings. This, she said, would amount to three months' work in all. Ayre had reason to believe that his Bristol employer's business was in a bad way, and decided to accept Rossiter's offer. His only previous contact with the Society had been as a visitor to the 1903 Show in Bristol, but he had no apprehensions, feeling that, as he put it, 'the work was my destiny'.

The 1910 Show was at Rochester, and he worked here for the whole period, February to September, while the showground was being prepared. During this time Rossiter became ill, but recovered sufficiently to see the job through. In Cardiff the following year he fell ill again, and died. Ayre was engaged by the Society in his place, on a temporary basis to begin with and then, 'the Council being satisfied by the way in which he carried out the trust reposed in him', permanently.

From then on, he was away from home continuously for seven months of the year. The only time his family went with him was in 1913, to Truro. The experiment was not a success, and thereafter, for the whole of his working life with the Society, his daughter recalled,[13] his family 'just didn't exist'. He rarely wrote, and never came home before the end of September. It was, Mrs Ayre used to say, 'like being the wife of a ship's captain in the sailing-ship days'. He was away when his son was born, away for his daughter's wedding and away when his wife had an operation for cancer.

He very seldom drew up a plan before he reached the ground, but he was a great believer in a show-site being square, to reduce walking to a minimum.[14] Despite frequent arguments to the contrary, he remained convinced that it did grassland good to have a show on it now and again. It improved the grazing, he believed, and reduced the weeds. In a wet summer, however, there was always a certain amount of damage and when the Show was held one year on the Downs in Bristol a very fussy Parks Supervisor insisted that every bare patch should be re-sown with a matching mixture. Ayre had the grass analysed and obtained appropriate seed, but he insisted that the Parks Supervisor should actually carry out the sowing himself.

The legal agreements between the Society and the owners of the Show site have nearly all disappeared. One does, however, survive for the year 1903, for Bristol. This document, written in the beautiful legal copper-plate that was normal at the time, reads:

'It is hereby agreed by and between the Lord Mayor Aldermen and Burgesses of the City of Bristol acting by Edmund Judkin Taylor the Town Clerk of the said City and The Bath and West of England and Southern Counties Society acting by Thomas F. Plowman the Secretary of the said Society and their duly authorised Agent in manner following (that is to say):

1. The said Lord Mayor Aldermen and Burgesses will raise no objection to the said Society holding their Annual Exhibition for the year One Thousand nine hundred and three on the same portion of the Clifton Downs Bristol (hereafter referred to as "the site") as was used by the said Society for their Exhibition for the year One thousand eight hundred and eighty six which site is shown on the plan submitted on behalf of the said Society and for the purpose of preparing for and holding such exhibition for the said year One thousand nine hundred and three will not object to the said Society having the use of the said site from the first day of February in that year to the following thirty first day of July but subject to the conditions hereinafter contained.

2. In consideration of such concession as aforesaid the said Society will:

    (a) Make all erections of the same description and construction as in past Shows of the Society and erect them so as not to damage any tree or shrub growing on the site and any suggestion by the Engineer for the time being of the said City on the subject shall receive due attention

    (b) Execute all necessary blasting operations under the direction of the said Engineer (but at the risk of the said Society)

    (c) Leave such openings in the hoarding to be erected by the said Society around the site as in the opinion of the said Engineer shall be necessary to permit of the public having free access and regress to and from the site at all times except for one continuous period of four weeks (to include the time of the Exhibition being held)

    (d) Preserve from injury all trees and shrubs growing on the site and on such other portions of the said Downs as shall be used as approaches thereto, and

    (e) At the close of the said Exhibition and to the reasonable approval of the said Engineer fill up all holes made for posts or other erections. And make good all injury caused to the surface of the said site and approaches and remove all litter broken glass and other rubbish therefrom

Dated the fifth day of June One thousand nine hundred and two'

It bears the signatures of the Lord Mayor, Charles E. S. Gardner, and of Edmund J. Taylor.

Every site brought its special problems[15] and its own characters to be dealt with. Mr Ayre always remembered, with both pleasure and respect, Mrs Morrell, on whose land at Headington Hall the Oxford Show was held. 'A nice person,' he recalled, but a lady resolutely opposed to the slightest amount of tree-pruning. There were, however, a number of trees on the showground with dead and dying branches, and Ayre had these dealt with while Mrs Morrell was away on holiday. When she returned, she liked the effect so much that she had all the trees along her much-overgrown entrance drive dealt with in the same way. 'You've done me a real good turn,' the head gardener told Ayre.

He had eight regular workers, who came away with him each year and shared his strange life. Outstanding among them were members of the Aish family, of Nailsea, the first of whom he inherited from Rossiter. Three generations of this family worked for the Society.

Until late March or early April, the Superintendent knew only the kind of shedding that would be needed. After that he was given the details of the exhibitors and the space each required. No exhibitor, he insisted, ever tried to bribe him to obtain a better position, but the occasional contractor, for such facilities as the water

46. Bristol, 1903. C. S. Bailey's hardware stand.

supply, offered him a tip. Complaints, however, were not infrequent. At Cardiff in 1911 two exhibitors made a fuss about the stands they had been allocated, alleging that their business would be ruined. Trade, as Ayre told them, would hardly have been good, no matter what position has been given to them. It was a hot summer, and they were both attempting to sell umbrellas and raincoats.

There were 150 tons of permanent plant, which had to be shifted each year from one showground to the next, but new shedding was built each year. The timber was sold by auction after the Show, usually for about three-quarters of the original cost. For the 1912 Show in Bath, Ayre was told that he would have to provide covered parking for cars—open sheds with a canvas roof. The space required was 200 feet by 20 feet, sufficient for twenty cars. One of the motors accommodated in this way in 1912 belonged to Mr Eaves, the Clifton photographer. No parking was requested the following year in Truro.

The Show had its good years and its bad years, the really bad years being due mainly to the weather. Rain was always anticipated—the English summer is notoriously unreliable—and early in his career as Secretary Thomas Plowman referred, with good reason, to 'the pluvial traditions of the Society'.[16] One of the few defences against unkind weather was a programme of indoor activities. A popular innovation of this kind was the annual Nature Study Exhibition. The first of these exhibitions was in 1903 and in 1905 at Nottingham, when the

47. Truro, 1913: arts and crafts exhibit.

organisation was well established, the exhibits consisted of 'zoological, botanical, geological, and mineralogical specimens and collections; illustrations (such as drawings, photographs, plans, diagrams, etc.); and models, appliances, apparatus, etc. illustrating the life-histories of animals and plants and the physical and geological features of the country.'[17] Invitations to exhibit were confined to educational establishments and private individuals within the region where the Show was held. In 1905 this meant the counties of Nottingham, Leicester, Derby, Lincoln, Rutland and the West Riding of Yorkshire.

The attendance at Nottingham was unfortunately 'not so large as had been anticipated, mainly owing to a great counter-attraction, viz. the first Australian Test Cricket Match. This was played at Nottingham during the Show week, and was attended by many thousands of spectators who would otherwise have found their way to the Show Yard.'[18]

In choosing where to hold the Show each year, the Society found itself constantly faced with a major problem. 'The Council,' reported the Secretary, 'have had under serious consideration the position in which the Society finds itself owing to the Royal Agricultural Society coming at short intervals into the area embraced by the Bath & West Society. This places the latter Society at a considerable disadvantage with respect to both its present and future, inasmuch as the cities or towns visited by the Royal Society are among the comparatively few at which the Bath Society can hold its annual Show without risk of a drain on its funds. Your Council would, therefore, at this critical juncture, earnestly urge all members of the Society to give it all the support they can, and to induce others to do so, so that it may be enabled to continue unimpaired the work which, for nearly one hundred and fifty years, it has successfully carried on for the benefit of Agriculture.'[19]

It was not that the Society was in a bankrupt condition. Between 1900 and 1914 there was a steady increase in its invested capital, and bank deposits remained satisfactory. The real trouble was that the subscription was no higher than it had been a century earlier, for ordinary members 'not less than £1' and for tenant farmers 'not less than 10/–'. And the records show that most members paid the minimum subscription. However well run a Society might be, and the Bath & West Society was very well run,[20] it simply could not pursue an adventurous policy without the necessary income. The Show might attract a lot of people, the *Journal* might publish excellent articles, the officers might be worthy and respected men, but if there was no money to finance a programme of experiment and research what, it might fairly be asked, was the Society really for?

One answer that was often considered, but not often put into words, was: to act as an intermediary between the Government and the farmers. In 1904, for instance, the Society had appointed a deputation to attend the Board of Agriculture to urge legislation to deal with warble fly and with contagious abor-

48. Truro, 1913: general view, including the Railway Companies' stands.

tion in cattle. With the outbreak of war in 1914, this liaison-role became inevitable.

The 1915 Show, at Worcester, was held as planned, but without the Butter, Poultry and Shoeing classes, and the Nature Study and Forestry Sections. The decision to go ahead with the Show was explained by the Secretary in these words: 'When in the opening days of August, 1914,' he began, 'the thunderbolt forged by the Kaiser descended upon a peaceful world, the Society was making

the necessary arrangements for holding its 1915 Show, as previously agreed upon, at Worcester.

'The natural and immediate effect of the declaration of War was to distract the
attention of the whole country from everything else. The war-cloud burst with
startling suddenness, for, so unsuspicious had the blessings of peace rendered
most of us, that only a comparatively limited few had the prescience to read
previous portents aright. When the crash came its effect was overwhelming in a

49. Truro, 1913: exhibition of grasses.

national sense, and every mind was bent upon considering, to the exclusion of every other topic, how the sudden emergency could best be met. Agricultural Shows at this moment shared the fate of other institutions outside the area of the War, and any thought for them was relegated to a time of less stress and strain in other directions.

'In the course of a few weeks it began to be realised that, whether the War was destined to be long or short, there was no immediate danger of the country being defeated by force of arms, or—thanks to the British fleet—starved into submission. Men's minds then began to be directed to the question as to how far the current of ordinary life should be deflected on account of the War, and those responsible for the conduct of the Society had to consider whether or not, under prevailing circumstances, it was either practicable or desirable to hold a Show in 1915.'[21]

Unusual and formidable problems had to be faced. 'For instance, not many days before the Show opened, the Hay Contractor conveyed the startling information that he had been formally served with a notice not to part with hay to anyone but the Government, and so he did not see how, under these circumstances, he could fulfil his contract with us. As every other contractor was in a similar boat, the prospect of having several hundred animals on one's hands with no provender for them was something appalling. However, a trustful belief in Providence and in the reasonableness of even a Government Department, when matters were fully explained to the latter, was not misplaced. The military powers that be, as represented by a very courteous official in charge of the district, saw the position we were in and not only removed the embargo but, in other ways, facilitated the acquisition of what we wanted. So let this be put down to the credit of the War Office in the days of its disparagement.'

All kinds of materials which had previously been taken for granted were suddenly not to be had. Old railway sleepers were a case in point. Each year the Society had bought a large number of them to make roads through the Showground. In 1915, and for the rest of the War, the Army commandeered the whole supply.

The Show was used as a recruiting centre for the Army. 'The recruiting-sergeants,' the Secretary reported, 'were busy in all directions, for the Society offered every facility for this. Those who were told off for this duty were admitted free and provision was made in the yard for the medical examination of likely candidates for military service. Many placards appealed to all who could to join the ranks of the King's Army, and the officers and men engaged appeared to be well-satisfied with the results of their efforts.'

Wounded soldiers were to be seen in their dozens on the Showground, 'limping about, or with arms in slings, or heads in bandages'. There were a number of military hospitals in and around Worcester, and patients from them were admitted free to the Show and given reserved seats on the grandstand and elsewhere.

Worcester was very much a war Show. The Office of Works had a pavilion, where farmers who were short of labour could call and get help and advice. A number of implement firms had 'exhibits designed to meet with emergencies' and the Board of Agriculture showed 'a collection of dried vegetables, such as are being sent out to the Expeditionary Forces'.[22] The Society could claim to have shown considerable foresight in the matter of dried vegetables, since it had been trying to encourage the development of this new industry for nearly twenty years. In 1902 the *Journal* had published a note by James Harper, in which he said 'In all high-class restaurants there is a steadily increasing demand for dried vegetables made up in the form of soups. Our Army in South Africa is now largely supplied with dried vegetables from Germany. There is no reason why such vegetables should not be dried in this country, and if, as seems not unlikely, such produce is utilised in time of peace as well as war for our Army and Navy, there will be a very large and increasing demand for it.'[23] But it took a major war to drive the lesson home.

No further Shows were held during the War—1915 was Thomas Plowman's last Show—and the equipment remained more or less safely in store at

Worcester. But the Society continued to function as a centre of information and to help both farmers and the Government in any way it could. The *Journal* continued to be published throughout the war, and a series of practical pamphlets was produced, some containing material reprinted from the *Journal* and others specially written for the purpose. These pamphlets, which were forerunners of the publications of the Ministry of Agriculture after the war, were authoritative, well-written and cheap, and found a market which the *Journal* rarely touched.

A list of those which had appeared by the time the war ended in 1918 shows the wide range of subjects covered. At a shilling each there were:

Observations on Cheddar Cheese-making, 1891 to 1898 (8 nos.)   By F. J. Lloyd, FCS, FIC

Investigations into the Manufacture of Cider, 1894 to 1900 (7 nos.)   By F. J. Lloyd, FCS, FIC

The Art of Butter-Making   By the Society's Dairy School Teachers

The sixpenny list was much longer:

The Farm Schools of Normandy and Brittany   By G. E. Lloyd-Baker

Permanent Pastures   By W. Carruthers, FRS

The Manufacture of Cheddar Cheese   By Edith J. Cannon and F. J. Lloyd, FCS, FIC

The Development of Collegiate Centres for Agricultural Instruction   By Douglas A. Gilchrist, BSc

Petty Industries and the Land   By J. L. Green, FSS

The Breeding of Light Horses   By John Hill

The Polo Pony   By John Hill

The Composition and Agricultural Value of Couch and Couch Ashes   By John Hughes, FIC

The Composition and Properties of Milk   By F. J. Lloyd, FCS, FIC

Organised Cottage Poultry Keeping   By Geo. F. C. Pyper

Construction of Dairy Herds   By Professor J. P. Sheldon

Dairying in New Zealand   By A. F. Somerville

Bovine Tuberculosis and its Suppression   By Jas. Wilson, MA, BSc

How Stock Breeding is Aided in Germany   By Tom E. Sedgwick

Both the 1914–18 and 1939–45 wars were periods of national stocktaking, in which past mistakes were assessed and regretted and proposals made to create an England fit for heroes to live in. The Society played its part in these philosophical adventures. In 1916 it published[24] an article by S. L. Bensusan on the conditions of the farm-labourer and the improvements that should be expected once the war had ended.

'The Labourer has been taught by the war to recognise his own proper value, possibly to overestimate it,' wrote Mr Bensusan, 'but this fault will right itself. He will have a measure of self-reliance and initiative to which he has hitherto been stranger. As a class he may be trusted to take the best advantage of a situation that has turned the tide of misfortune that began to submerge him when, in the last great struggle for national existence a hundred years ago, the commons were enclosed and the yeomen dispossessed. Nothing less than this tragic European upheaval would have availed to alter his position so suddenly. Emigration would have done it in time, the pace of progress was steadily increasing, but that was evolution and this is revolution. Under normal conditions there would always have been an inexperienced and timorous residue content to endure the known rather than face the unknown evils. The ranks of agricultural labour have not produced half a dozen great men in a century, the work exhausts the body at the expense of the mind, and it is in teaching the farm hand to think that the drill sergeant has laid the foundations of a peaceful upheaval. Will those who return prove able to instil the spirit of initiative into those they left behind? Will this spirit be strong enough to impose upon farmers as a class a sense of the necessity of putting their house in order? It is dangerous to prophesy, but to those who know the farm labourer and his problems, who recognise his merits as well as his short-comings, and who feel that the future owes great atonement for the past, the years that directly follow the end of the present war will be as full of interest as of hope.'

Ten years earlier,[25] Sir C. T. D. Acland had seen the situation less optimistically and, uninspired by the wartime atmosphere, possibly more realistically. 'We must take for granted,' he believed, 'that, as things are at present, the brightest and most teachable boys in the country schools are not likely to remain in the rural districts as agricultural labourers'. Those who would be trained for farm work were likely, for the most part, to be 'mediocre children coming from uneducated homes and very illiterate parents'. The future would be 'one in which their employers will require intelligent, observant, upright and teachable servants. We must do our best to render them also happy and contented, though not unenterprising'.

Acland, like his father before him, took a keen interest in agricultural education at all levels. After his death in 1918, the obituary, written by Plowman,[26] drew particular attention to this. 'Before the Board of Agriculture and County Councils came into being, little help was afforded to the most important of our industries beyond that rendered by agricultural associations. This induced the Society's Council to set on foot an important scheme to provide for systematised investigations and practical demonstrations in connection with the various departments of husbandry and for establishing experiment stations and educational centres for the testing and promulgating of improved methods in relation, especially, to dairying. Mr Acland was one of the most active supporters of this new departure and rendered essential aid in bringing it to a successful issue. During the years in which the scheme was in operation, it had an important bearing upon agricultural education, as its subsequent recognition by the State testified. It was only relinquished when the establishment of the Board of Agriculture and other public bodies, either State-aided or Rate-aided, rendered the Society's assistance in the directions indicated superfluous.'[27]

In arguments over policy, Plowman recalled, Acland had been forced to give way only once. This was over the matter of money prizes for stock at the Show, 'his view being that the honour and glory of winning, coupled with its value as an advertisement, and with the addition of a medal as a memento, should be all-sufficient to satisfy exhibitors. With indomitable perseverance, he endeavoured to convert the Society's Council to his views, but it was one of the rare occasions when his colleagues failed to follow his reasoning. The exhibitors, too, manifested a distinct preference for hard cash.'

Plowman himself died in the following year. The Society's debt to him was incalculable, not least for his great efforts on its behalf during the war years, when the difficulties that had to be faced would have tried the patience and the stamina of a man fifty years younger. His only personal regret was that, for the last three years of his life, there could be no Show. The Great Showman had to go out without a Show.

1919 and 1920 were a fool's paradise. At the end of 1920 wholesale prices were 220 per cent above the 1914 level, wages 180 per cent and the cost of living 160 per cent. Huge profits were made and spending, after the years of war-time shortages and restrictions, ran wild.

The collapse in the national economy was rapid and brutal. By March 1921 the number of registered unemployed had risen to 1,664,000 and at the end of the year it was just under two million. Between then and the outbreak of war in 1939 the total rarely fell below a million and a half. In September 1931, it was nearly 2,900,000 or more than one in five of the labour force. The General Strike of 1926 expressed the despair and anger of men in employment who found it impossible to maintain a tolerable standard of living on the wages they were paid.

Agriculture inevitably shared the general depression. World food prices were too low to give producers an adequate return on their investment and a great many farmers, in Britain as in other countries, went bankrupt or sold up their farms. The English countryside took on an increasingly neglected look and remarkable efficiency and enterprise was needed to make any profit at all at the foodstuff prices which had to be accepted. The urban population bought the cheapest food it could get, and was in no position to support British farmers simply because they were British.

It would be wrong to give the impression that the plight of rural England was a matter of indifference to the Governments of the day. This was far from being the case. A mass of legislation was enacted by successive Governments in an attempt to arrest the diminution of the arable acreage, or at least to alleviate the general depression in the industry. These measures included the Wheat Quota Act; Legislation to extend and subsidise the sugar beet industry; a number of Marketing Acts; the establishment of the Milk Marketing Board; and several Rural Housing Reconditioning Acts.

By comparison with manufacturing industry, agriculture had relatively little unemployment, but the level of wages remained considerably below the national average, and it was difficult to persuade younger men with intelligence and ambition to have anything to do with farming.

One more cheerful feature of this dismal period should not be overlooked. During the 1930s it was possible to buy farms ridiculously cheaply and a number of people with non-agricultural money to spend took advantage of the opportunity, reckoning that prices could hardly be lower and that, by bringing this neglected land back into order, an interesting task in itself, the investment might eventually be rewarding. Men and companies who thought like this had good reason to congratulate themselves during the '40s and '50s, one of British agriculture's boom periods of all time.

Depression and dullness between the wars

Plowman's letter of resignation was dated January 16, 1919. In it he said that he felt unable to run a Show when this activity was resumed by the Society. 'You might be reluctant to convey to me,' he wrote, 'what I have taken upon myself to say.'[1] A special committee was set up to consider the matter. It met in March, and decided that the salary to be offered to the new Secretary would be £750 a year,[2] out of which he would be expected to meet the cost of any office staff, with the exception of the Assistant Secretary, W. A. Smith, who had a separate salary of £300.

In May 1919, the committee decided to consider two applications for the Secretaryship, out of nine received. The shortlisted candidates were Capt. F. H. Storr and J. Nugent Harris. Storr failed to appear at the interview, although he was clearly the favoured candidate, and it was agreed to appoint Harris if Storr should withdraw. In July, however, Storr presented himself for interview, and was given the job. He had private means and sufficiently good connections to overcome the competition for the post, despite his lack of obvious qualifications for it. The Council included the news in a general report on administrative changes. They had, they said, 'appointed as Secretary and Editor Mr F. H. Storr, OBE, who during the latter stages of the War had been acting as Intelligence Officer on the General Staff, and who was strongly recommended by those with whom he had previously been connected. Further, in view of his long term of service with the late Secretary as Chief Clerk and the value to the Society of his special knowledge, they had appointed Mr W. A. Smith as Assistant Secretary.

'In making these appointments your Council felt that the responsibility for the publication of the Society's Annual Journal should rest entirely with the Secretary and Editor, and they have, therefore, terminated the appointment of Associate Editor held by Mr F. J. Lloyd since 1890. In doing so, they desire to record the Society's indebtedness to Mr Lloyd for the valuable services he rendered the Society during that period.'

The contents of the first volume of the *Journal*[3] to be produced under peace-time conditions illustrated the changes which had taken place in agriculture and its associated industries during the war period. There was an article on the Ministry of Agriculture's milk recording scheme, introduced in 1915, and others on tractors and on the pollution of rivers by milk factories. None of these topics had attracted the Society's attention previously. They were, so to speak, war-time bonuses.

Tractors were a response to the shortage of manpower, and to the need to bring a larger acreage under cultivation. In Britain, said H. E. Gardner,

'Saundersons, of Bedford, probably, were the pioneers in farm tractors. In America, the Hart-Parr Company claim the distinction. In general, developments have proceeded with the advance of motor engineering, and within the last four years this country has witnessed many developments in farm tractors, so that to-day what is practically a new industry has been established.'

The extent of river pollution was only just beginning to be recognised for the serious problem that it was. 'Year by year,' declared the *Journal*, 'this pollution has been increasing, until at the present time it has become such a very serious matter, that Local Authorities have at last been compelled to pay attention to it. Proprietors of these Factories have certain obligations, under the Rivers Pollution Acts, which are clearly defined. In the case of industrial factories in existence before the passing of these Acts, effluents must not be discharged into streams unless reasonable and practically available means have been adopted for rendering them harmless.'

In 1921[4] Professor James Long took stock of agriculture in Britain, after several years' intensive activity on the part of the Government to encourage and bribe farmers to produce more food. His article, 'Farming and Reconstruction', painted a gloomy picture. 'It is impossible to ignore the fact,' he wrote, 'that British agriculture is in an unsatisfactory condition. It has been failing the country for many years.' The war, he believed, had not changed the situation in any fundamental way. Anyone who had travelled round the country with his eyes open was well aware that 'a very large proportion of the land under crop is only half cultivated or not cultivated at all, and that there is an enormous area of rough grazing land which ought to be brought under cultivation'.

The Society had already embarked[5] on a research programme of its own to see how poor pasture-land might be improved. Five demonstration plots were selected for long term experiments, to discover the best ways of eradicating bracken and thorn bushes and to assess the results of applications of lime, basic slag and phosphate.

Work of this nature was not costly—the days were long past when even such a large society as the Bath & West could organise and finance anything other than minor research projects—but it was valuable to have it carried out at a number of centres well distributed over the country, so that as many farmers as possible could visit the demonstration plots and see the results of the experiments for themselves. Large-scale research had become the business of specialised insitutions, which had the advantages of Government money and full-time expert staff.

Within the Society's area, Bristol University's Agricultural Research Station was felt to have the greatest claim on its attention and support. Founded at Ashton Court in 1904, as the National Fruit and Cider Institute, it subsequently became part of the University and was generally known as Long Ashton

Research Station. From the beginning, the Society supported it with a grant of £100 a year, nominated the very able Henry Burroughes Napier as its representative on the Governing Body and published its Annual Report as part of the *Journal*. As the work of the Institute grew, these Reports grew longer and more expensive to produce. In 1920–21 ninety-five pages were required, in 1921–22 one hundred and thirty-two pages, and in 1923–24 one hundred and fourteen pages. By then the Society had decided, not unreasonably, that its generosity had become excessive, and that some limitation was desirable for the future. Announcing this[6] it pointed out that the scope of the Research Station's work had greatly increased and that 'the technical character of many of the details it is now necessary to record, though of great scientific interest, are not always capable of immediate practical application'. It was therefore arranged that 'while a general survey of the work at Long Ashton will still form part of the report made to the Society, only those subjects which are of general interest and on which fairly definite conclusions have been arrived at, shall in future be published in the Journal. Reference must also be made to the valuable educational exhibit made by the Research Station at the Society's meetings, which alone more than justifies the continuation of the grant.'

The Shows of the early '20s were dogged with exceptionally bad luck. In 1923 the Stewards had to deal 'with an unprecedented difficulty, and one which threatened the success of the Show at its start, namely, the failure of the railway company concerned to arrange in the least adequately for the stock exhibits on their arrival. In spite of the most urgent representations many animals were kept for more than 15 hours in the railway sidings, and arrived on the Show Ground so late that judging had to be postponed. The representative of the railway company expressed his deep regret for what had transpired; promised that a thorough investigation should be held, and that adequate arrangements should be made for the outward traffic. He expressed a hope that what had occurred would not prejudice the goodwill of the railway, and agreed that their apology should be read to the Annual General Meeting of the Society.'[7]

1926 was even worse. 'It is not easy to imagine an Agricultural Show held under greater difficulties than those which confronted the Society in 1926 at Watford,' reported the Secretary.[8] 'The Show was due to open on May 25th, and on May 3rd the General Strike broke out so that it was not until 9 days before the opening that it was possible to decide that the Show need not be abandoned. In fact, even at that date, the Railway Companies were dubious of their ability to make the necessary arrangements. Transport was not the only important department affected. The stoppage in the printing trade, the effects of which lingered after the general strike was at an end, made it difficult to produce the Catalogue, and, in addition, interfered seriously with the production and distribution of posters advertising the Show. Those posters, too, which were displayed in the immediate neighbourhood of Watford were not in favourable positions. However, the Railway Companies and the printers rose to the occasion, and the Show when opened showed very few signs of the difficulties under which it had been completed. But nothing could make up for the lack of facilities for passenger traffic on the railways, where, instead of a good service of trains at cheap fares, only half the ordinary trains were running at the ordinary charges.'

The situation was back to normal the following year, which marked the 150th anniversary of the Society. The Show was at Bath,[9] and 'the laying out of the Show Yard was universally pronounced to approach very nearly to perfection, and reflected the greatest credit on the Superintendent of Works, Mr H. C. Ayre.'[10]

On the third day of the Show, the Minister of Agriculture, Col. the Hon. Walter Guinness, unveiled a memorial tablet to Edmund Rack at 5 St James's Parade, Bath, which had been Rack's home. In his address the Minister pointed out that the Agricultural Societies in Yorkshire and Norfolk, from which Rack took his inspiration, had not lasted long, but that the foundations of the Bath Society had been 'laid on surer foundations'. The Minister said that he 'felt it a privilege to take part in commemorating a man who devoted his brilliant talents for the advantage of agriculture, and of his fellows'.

This Show made a special feature of what it called 'those minor industries and activities of the countryside, which it is of growing importance to maintain and increase'. It believed that it made a greater effort than any other society to encourage people to take an interest in the traditional crafts and in the history of the countryside, and for its 150th anniversary a very enterprising exhibition was got together, with the assistance of the Federation of Women's Institutes, the British Legion and the Somerset Rural Community Council, none of which, incidentally, had existed before the War.

Part of this exhibition was devoted to 'Old Agricultural and Farmhouse Implements'. Large crowds came to see it, 'and much interest was taken in the exhibits, while Mr Martineau's three-minute lectures on the Ancient eight-ox Plough, and its influence on the English land-measurement, were very well received, particularly by technical experts, who made many useful comments'.

The range of implements and equipment was quite wide. 'Mr Neville Grenville showed a hand turnip-cutter, some harvesters' cider barrels, and Sir Frederick Bathurst lent a very fine flail with the usual ash staff and thorn beater and a curious holly-wood swivel at the joint. Mr A. F. Luttrell, of Dunster, showed four ox-yokes, Mr W. Pullen, a set of horse bells and ox-bell, a flail, ox-cues and other things, Mr H. G. White a fleam and mallet, Mr G. Drew a hummeller—and it is worth noting that five other names were collected for this implement from visitors. They are, (1) Awner, (2) Aveller, (3) Chumper, (4) Stumper, and (5) Spiler, the last three being local. Aveller is the name in the Catalogue of the Great

Exhibition of 1851. The Lord Wandsworth Agricultural College showed a model of the original hop-bagging machine; Mr Rawlence, a series of horse shoes of various ages, with bullock shoes and other small ironwork, but the Show would have been small had not Mr Lavington of Bathford kindly lent a large number of domestic utensils and a few obsolete tools. The old Kentish plough, lent by Messrs. Drake and Fletcher, excited much interest and served as the text of Mr Martineau's discourses. This plough is pictured in the Society's *Journal* for 1786.'

It had become clear by this time that the main activity was to be the Show, which had become not so much the Society's shop-window as the reason for its existence. This was partly because few other outlets for its energies appeared to remain, after research and education had been taken over by local authorities, universities and Government institutions, and partly because a successful Show helped the Society's prestige and brought members together in a way that nothing else would.

1928 at Dorchester was a Show much above the average. The Duke of York attended and the attendance figures were exceptionally good as a result. This was the first occasion, too, when the motor-car really came into its own. 'The sight of several acres covered by some hundreds of cars was distinctly impressive,' reported the Secretary,[11] 'and completed the effect produced by the admirable decorations in the main streets of the town.'

50. Dorchester, 1928: visit of the Duke and Duchess of York.

Another noteworthy feature of the Dorchester Show was 'a demonstration by sheep dogs'. Great interest was taken in this, 'as was natural in a county where sheep are of such importance'. The shepherd in charge of the demonstration was 'Mr J. B. Bagshaw, the winner on three consecutive occasions at Llangollen, in many ways the most important of the big Sheep Dog Competitions, and the English champion in 1926. He brought two young dogs, Cap and Moss, the exactitude of whose response to direction was astonishing. Capt. W. Best, to whose initiative the holding of the demonstrations was due, sent down some Welsh mountain sheep for the purpose, a necessary precaution in a locality where sheep are not generally used to trained dogs, and the performances given were a great success.'

The 1929 Show, at Swindon, was the first four-day Show the Society had held. It was difficult to decide if the experiment had been a success or not. 'In the first place,' the Secretary explained, 'the date of the Show had to be put forward by a week under circumstances explained in the Annual Report of Council with the result that the Show immediately followed the Whitsun holidays, a great disadvantage, above all in an industrial town such as Swindon. But a still greater handicap was the continuance of the epidemic of Small-pox in the town, advertised by the action of the Southern Command in putting Swindon out of bounds for their troops. When account is also taken of the imminence of the General Election, it will be seen that no safe conclusions can be drawn as to the effect of a four days as against a five days Show.'[12]

The Prince of Wales paid a visit to the show in 1930, at Torquay. He arrived by air, and it was noted that 'the speediness of air communications enabled him to combine with his visit the inspection not only of a scheme for extending the water supply of Torquay, but also of the more westerly of his Duchy property. Duchy tenants from a wide area were given the opportunity of meeting His Royal Highness in the Show ground, and it is hoped that this popular function may be repeated at future Shows of the Bath & West.'[13]

But, despite the occasional profitable bonus, such as the Prince of Wales descending from the clouds, the Annual Show was always a great worry. The Society's view continued to be that there were far too many shows each year, and that with motor transport now so much more widely available, the number could be considerably reduced, to the benefit of the public, the exhibitors and the societies. But each society was bound to reckon that without a Show there was no point in its existence. The only solution, therefore, was to amalgamate societies. In 1929–30 there were negotiations for a merger between the Bath & West and the Royal Counties Societies, but the plan was eventually turned down by the Council of the Royal Counties. In 1936–37 another attempt was made and this time, although the Royal Counties Council was in favour, its members, by a small majority, failed to give the scheme their support. Having reached a

51. The Prince of Wales at the Torquay Show, 1930.

dead end in this direction, a new form of co-operation was worked out with the
Royal Counties and Three Counties Societies. For some years demonstrations
of agricultural implements had been organised by County Council officials, and
these had been found to be of considerable interest and value to farmers. These
ad hoc arrangements were put on a more permanent and better co-ordinated
basis, under the title of 'The South of England Agricultural Demonstration Committee'. The Council of the Bath & West agreed to appoint five representatives to
the Committee, and to grant it the use of the Society's offices and staff. The plan,
as announced in 1937, was to hold a Spring and Autumn demonstration each
year in an area covered by the counties of Hampshire, Wiltshire, Dorset,
Somerset and Gloucestershire. The Committee, it was announced, 'has the
backing of the Ministry of Agriculture, the University of Bristol, the Royal
Agricultural College, Cirencester, and the County Councils concerned.'[14]

To begin with, the scheme operated very much as it was intended to do, but it
had less than two years in which to operate before the Second World War broke
out. It was never revived.

The Society had more success with the Wiltshire Agricultural Association,
which was taken over in 1932. The effect, however, was to remove competition,
rather than to add strength to the surviving Society. Membership in 1933 was
almost exactly what it had been in 1932. This, even so, the Society thought,
should 'be considered a satisfactory state of affairs in view of the considerable

losses shown by the membership of some other societies, and of the financial stringency in the country'.[15]

The general gloom produced by what the Society described as 'financial stringency' and other people 'the Slump' had been lightened a little by an important decision of the Court of Appeal in 1927 that agricultural societies should be considered benevolent institutions and, as such, not liable to income tax. The cost of the litigation needed to achieve this result had been heavy. It had been shouldered principally by the Bath & West, the Yorkshire and the Royal Lancashire Societies.

The Society was not over-blessed with energetic, imaginative people during the Twenties and Thirties, and the loss of H. B. Napier in 1932 was consequently particularly sad. A land agent by profession, he accepted the position of Agent to Lady Smyth, of Ashton Court, in 1900 and held this post under her successor, the Hon. Mrs Smyth, until he died at the age of 76. In 1888 he was elected a member of the Society's Council and within the next four years he was appointed Steward of Works and a member of the Agricultural Education, Allotment, Contracts and Experiments Committees. He became one of the Governors of the first representative body of the National Fruit and Cider Institute in 1904 as a nominee of the Society. He was at once elected a member of the Institute's Management Committee and became Chairman of the Governors in 1906. He took a leading part in the negotiations leading up to the association of the Institute with the University of Bristol[16] and the establishment of the Department of Agriculture and Horticulture in the University, with its headquarters at what then became Long Ashton Research Station. He was appointed Chairman of the Sub-Committee of the Department, and he was also Chairman of the Sub-Committee which had charge of the Fruit and Vegetable Preservation Research Station at Chipping Campden. Within the Society, he became Chairman of the Contracts Committee in 1908, Chairman of the Finance Committee in 1911 and a Vice-President in 1922.

Napier illustrates admirably what one might describe as the public-service element of the Bath & West, its support for worthy bodies and solidly established research organisations. But it is possible to suggest that, during the 1930s especially, the Society's concept of public service was becoming rather old-fashioned. This was, after all, a period during which millions were unemployed and the countryside, like the industrial areas, was stagnant, if not rotting. One looks in vain in the *Journal* and the Minutes for any traces of crusading spirit, of willingness to take political action to break through the defeatism and fatalism which made this last decade before the Second World War so depressing and lacking in initiative. Committees were held, the Annual Show took place with faithful regularity, old Vice-Presidents died and new ones were elected, but there was little evidence of any wish or capacity to be a spearhead of innovation. Such novelty and change as there was tended to be very much on the surface. On the third day of the 1935 Show, for instance, 'the BBC broadcasted in their Western Regional Programme a discussion on the Show in which the Society, the visitors, and the agricultural experts were represented. Appropriate noises of machinery and the clicking of turnstiles were successfully interpolated, even the animals from the stock lines were induced to say their part'.[17] The Show, thanks to the BBC, was certainly finding a wider public than ever before, but the Society, as it had been reminded more than once during the nineteenth century, does not live by Shows alone.

The *Journal* during the last ten years of its existence—one hesitates to say life—was not an exciting publication. Dull outside, it was hardly more interesting inside. A long article, in 1937, on the cultivation and uses of the fuller's teazle, might be classified, according to taste, as either marginal or escapist. It was hardly evidence of a vigorous editorial policy, relevant to the current situation of agriculture.

The new Secretary, J. G. Yardley, fresh from the Royal Highland, where he had been Assistant Secretary, took over a Society characterised by an outstanding reputation and excellent potential, but low-profile present performance. 'I came and took a chance,' he said.[18] 'There was a good Council, and they gave me a free hand.' He made no secret of his determination to pull the Society out of a rut. His Show policy was simplicity itself. 'I wanted something to entertain the

52. Hackney class, Taunton, 1935.

53. The Society's Coronation Medal, 1937.   54. Ditto, reverse.

public,' he recalled,[19] and he set out to do just this. At Plymouth, just before the war, he achieved a high attendance, despite the restrictions and difficulties caused by foot-and-mouth disease. Yardley attributed this to good ring events, which he called 'Yardley's Circus', and to organising a special service of ships to bring Cornish people to the Showground from Torpoint. The 1939 Show at Bridgwater was also very successful.

But the membership refused to grow to the extent that the Council felt it should. This was plainly stated in 1939. 'Although over one hundred and fifty new Members had been elected during the current year,' Yardley reported, 'mainly due to the increase in the Non-Membership fees for Trade Stands, the total Membership is still unsatisfactory, and your Council are of the opinion that, with the growing influence of the Society, a much larger Membership is essential if it is to continue to carry out its part in the Agricultural life of the Country. The Council have had this matter under consideration and hope to bring before the next Annual General Meeting suggestions whereby the Membership of the Society may be increased. In the meantime, however, they trust that the present Members will interest themselves in this very important matter and endeavour to do their utmost to obtain new Members so that the Membership may, in the near future, reach a total more in keeping with the status of the Society.'

Two months later war broke out, and that, for the moment, was the end of all plans for growth and development. The Secretary went away to work for the Glamorgan War Agricultural Executive Committee and came back to Bath at weekends to keep an eye on what was happening there. During the week, the office in Pierrepont Street was looked after by Ayre, the former Superintendent of Works, and, with very little of any consequence to be done, this system worked quite well.

55. At the Plymouth Show, 1938.

By September 1939, arrangements had already been made for the publication of the 1939–40 volume of the *Journal*, but the Members of Council were asked for their opinion 'as to whether the *Journal* should be continued beyond 1940 in the event of war not terminating before the end of that year'. Council decided 'to cease publication of further volumes until the end of hostilities'.

In place of the *Journal*, the Society began to publish a series of practical pamphlets, as it had done during the 1914–18 War. These continued to appear until 1956, by which time there were twenty-four in all.[20] Authoritative, clearly written and sufficiently brief for a busy farmer to read, even in wartime, they were a thoroughly worthwhile project and performed a valuable service in keeping the name of the Society in front of the agricultural community at a time when other activities were at a standstill.

During the six years of war, British agriculture was transformed, at least temporarily, from a neglected, unwanted paddock into a cherished, subsidised, desperately needed meadow. Only half-believing that the days of depression and bankruptcy were over, farmers began to learn the benefits and the pleasures of bringing land back into cultivation, buying up-to-date equipment and modernising their premises. The banks and the Government smiled on them for the first time for many years, and they emerged from the war in a state, if not of affluence, at least of reasonable comfort and security.

Between 1939 and 1945, and indeed for a decade afterwards, the Ministry of Agriculture played a more active part in the British countryside than it had ever done before. Its local agents, the War Agricultural Executive Committees, kept a tight, but encouraging, hold on everything a farmer did, supervising his cropping, his feeding-stuffs, his labour and his sales. Very hardworking teams of experts demonstrated new techniques, persuaded and bullied farmers to adopt methods and working habits which would increase production and held their hands until they were convinced that such breaks with tradition as autumn calving were not going to spell ruin.

The rationing system and the great shortage of imported foodstuffs meant that the British farmer could sell, without any effort, everything he produced. It meant, too, that for the first time since the 1914–18 war the British public became aware that there was nothing automatic about food supplies, that bread, meat and butter from across the Atlantic did not buy and transport themselves, and that it was prudent to grow as much as possible in one's own country. Through the press, the radio and through his own stomach, the urban man-in-the-street became a little more aware that milk came from cows, butter came from milk, and corn grew in British fields.

During the late 1940s and the 1950s, rationing eased and then disappeared and the urgency of food-production slowly faded. Imported animal feeding-stuffs once again became easy to obtain and large intensive systems of meat and egg production became normal. There were fewer workers on the land each year and the increased wages they were paid caused farmers to substitute machinery for men wherever possible. Country people were coming to expect much the same amenities as towns-people, and the old rural community disappeared further and further into history with every year that passed.

The Bath & West necessarily reflected these changes, although, as a custodian of traditions as well as a midwife of progress, it was sometimes a little slow in approving and encouraging them. But it came to terms very quickly and willingly with what was perhaps the most significant social change of the post-war period, the mass ownership of motor-cars. Once this had occurred there was little point in taking the Show round the region to the people. The people could come to the Show.

A permanent home

With the war over, the Society decided that it would be impossible to return to the old system of building its own Show year by year. The task was entrusted to a firm of contractors, L. H. Woodhouse and Co., of Nottingham, on a series of three-year contracts. The new partnership worked very well. Woodhouse's showed great ingenuity in discovering timber during the immediate post-war period, when material of any kind was extremely hard to get. The 1947 Show fittings, at Cheltenham, contained wood from sources as unlikely as old bedsteads.

Cheltenham was, in view of conditions, a remarkable success—104,000 paid to visit it—but Cardiff, in 1948, beat all previous records. The attendance, 162,000, was the highest ever[1] and Yardley, who always considered Cardiff to have been his best Show, recalled with pardonable exaggeration, 'We couldn't cope with the money'.[2] The 120 acre site at Pontcanna Farm, off Cathedral Road, was ideally placed in the middle of the city. The Agricultural Engineers Association decided to permit their members to exhibit and there were so many applications for stands the front entrance had to be taken down and brought forward 1,500 feet to provide the room required. The Machinery-in-Motion section was one of the main features of the Show, with the biggest array of machinery and implements that the Bath & West had ever put on. As the Annual Report said, 'all the latest inventions and improvements were on view, and no one could fail to appreciate the efforts of the large Agricultural Machinery firms to assist the Ministry of Agriculture and the farmer in the campaign for increased food production'.[3]

There was an enormous Military Section, where the exhibits included Hitler's will, Hitler's telephone exchange, Goering's bullet-proof Mercedes and two staff cars used by Field Marshal Montgomery, 'Old Faithful' and the 'Victory Car'. The Royal Engineers demonstrated mine detectors, the RASC had its Mobile Machine Bakery Platoon, the Army Catering Corps showed the preparation and cooking of specimen dishes and the Household Cavalry put on a Musical Ride.[4]

The Educational Exhibits were given great prominence. There was a large section organised by the Ministry of Agriculture's National Agricultural Advisory Service, with a Working Dairy Laboratory, demonstrations of hay and silage-making and grass-drying, and displays showing the value of John Innes compost, the use of John Innes compost and the control of pests. 'The principal aim of the NAAS Exhibit,' the Catalogue explained, 'is to stress the importance of attaining a high degree of self-sufficiency in feeding-stuffs and to demonstrate modern methods of growing, conservation and utilisation of crops'.

On the livestock side, the NAAS showed how farmers could grade their cattle

198

up from commercial to pedigree standards, obtain more lambs and more mutton from better crosses, increase bacon supplies and improve the health of their livestock. The Welsh Plant Breeding Station presented information about new crop strains, the Institute of Agriculture for Monmouthshire explained the educational opportunities available for the sons and daughters of farmers, the Department of Agricultural Economics of the University of Wales explained the financial basis of a farm's operations, and the Forestry Commission and the Royal English Forestry Society combined to organise a four-acre exhibit centred on 'the great work being carried out in the restoration of the devastated woodlands caused by the demands of the nation during the war period'.

There was a large Horticultural Section, competitions of all kinds, Jumping, Sheep Dog demonstrations, and Parades of Foxhounds. On the second day, 'Her Royal Highness the Princess Elizabeth honoured the Society by an official visit, and was given a tremendous ovation whilst walking through the Showyard'. The 162,000 visitors got their money's worth and, by any standards, the Show was an extraordinary feat of organisation. In one way, however, the Society could hardly have gone wrong. After the long dreary war years, the British public was starved of big events like this, and it would have been a poor show that failed to attract them. The Council made the point slightly differently. 'The extraordinary interest shown by the general public in agricultural shows since the war has been most encouraging, and of great benefit to the Society,' it believed, 'and we must do our utmost to retain that interest in the future'.[5]

What was slightly bothering, however, was the fact that the Annual Show, however successful, was not bringing in the additional membership that was, as always, so badly needed. The published figures were not altogether what they seemed. In 1948 there were 2,750 members who had paid a subscription for the year. Of these, however, it was anticipated that 'a large number' would drop out, because they had enrolled themselves only to obtain Members' privileges for the current year's Show.[6] In an attempt to discourage this a new rule had been approved, providing that anybody who applied for membership between February 1 and the conclusion of the Annual Show held in that year should pay an additional entry fee of £1 in addition to the subscription for the year.

At the Annual Meeting at Cardiff it was announced that the Secretary of the Society would be taking over the Secretaryship of the Smithfield Club as well. The first post-war Smithfield Show was to be held in only a few months' time, on December 5–9, and that additional staff would have to be recruited in Bath in order to get through the work.[7] The existing full-time staff consisted of the Secretary, Assistant Secretary, Superintendent of Works, First Clerk, Second Clerk, two typists and one office boy. The Smithfield Club would be paying the Society £1,250 a year towards the cost of extra staff and the provision of office accommodation. But, as the Council pointed out, 'no money has been spent on the offices for a considerable number of years and they are badly in need of urgent attention'.[8]

By the time of the 1949 Show, the drive to obtain more members had begun to show results, and the total had risen to 3,253. This, however, was considered 'totally inadequate', because the Council was not content with running a successful Show and said it was 'very anxious to increase the usefulness of the Society in other directions—in the field of Agricultural education, the carrying out of experiments and the trials of new implements, besides giving every possible encouragement to the Agricultural Industry in the part it has to play in the nation's recovery programme.'[9]

It is fair to point out in this connection that the big money-maker was the Show. In 1949, the income from subscriptions and donations was £3,253 and from dividends and interest £2,045, but the Bristol Show yielded a profit of more than £25,000, an above-average figure, it is true, but even £10,000 would have made the Show the main bread-winner. On this occasion, the Council, its morale high, decided to publicise the fact that its aim was 'to knit more closely together the urban and Agricultural communities' and that 'to excite the interest of the townsmen it is necessary to stage at the Show attractions other than Agriculture, which have a distinct appeal to the ordinary person in the street.'[10] 'Yardley's Circus' might bring them in, but 'once in the Show it is hoped the townspeople will take the opportunity of seeing for themselves the great efforts of their country cousins to produce for them the highest quality food under the most hygienic conditions, and to realise more than they have in the past the many difficulties which confront the farmer and his servants in their daily tasks. In these drab days of frugality a little brightness is welcomed and keeping this in view the Society provided in the Main Ring attractions of very high quality including musical rides by the Household Cavalry, in full dress, to music by the Life Guards who also provided musical entertainment in the Band Stand in the centre of the Yard. Other items were counter marching by the Band of HM Royal Marines, Plymouth Division, also in full dress, sheep dog demonstrations and, through the kindness of the President, His Grace the Duke of Beaufort, MFH, and Colonel H. C. C. Batten, DSO, MFH, Parades of Fox Hounds of the Beaufort and Cattistock Hunts.'

There were also the horse jumping competitions, which were always a great favourite at the Show. At Bristol, 'the jumps were far more testing than those usually seen at Shows, and in some cases as many as fourteen jumps had to be negotiated, forming a fine test for horses and riders'.

Two much appreciated innovations at the 1949 Show were a livestock auction competition for members of Young Farmers' Clubs and a centre, staffed by the WVS and YWCA, where mothers could leave young children, while they looked round the Show in peace. This very sensible idea, which became a feature of future Shows, allowed many women to come who would otherwise have been prevented from doing so.

Too few visitors became Members, however, and in 1951 the Council had to report a decline in membership. This, it thought, was both disgraceful and incomprehensible. 'As the oldest Agricultural Society in the country, its membership compares unfavourably with that of the Royal Agricultural Society of England, and even some County Societies,' Members were reminded. 'For instance, the Yorkshire Agricultural Society has a membership of over 10,000. Surely the Bath & West, which covers a much wider area, should have a membership four times its present number.' Possibly without realising it, the Council had put its finger on the root of the problem. The Bath & West covered too large and heterogeneous an area to make it possible to appeal to local patriotism. The existence of its four Divisions,[11] each electing members to the Council, was in itself an admission that a slab of Britain stretching from the Medway to the coast of Wales was too vast and unwieldy to be considered as a whole. 'Yorkshire,' by contrast, meant something definite, something to be proud of. Nothing, it appeared, would move the membership total in the right direction. In 1951 there was an appeal to each member to obtain, before the end of the year, at least two new members. The result was not an increase but a slight decline.

Despite this apparent apathy, a limited programme of research and experiment continued. There was even a return to the old premium system, as a means of encouraging enterprise and invention. The great ditches, or rhynes, which drain the Central Somerset plain, were difficult and expensive to keep clear of weeds and in 1951 the Society offered the handsome prize of £1,000 for a Rhyne Clearing Machine. Inventors, as it turned out, were not equal to the challenge. Only two entries were received and these were tried out on a Sedgemoor farm. 'The type of rhynes selected were typical of the area,' said the report,[12] 'and were sufficiently covered with weeds to provide a severe test for the machines.' The judges decided, however, that neither machine did what was required of it, but the inventors were each given a grant 'towards the expenses they had incurred in sending their machines for trial'.

From a financial point of view, 1952 was a disastrous year.[13] A loss of more than £5,000 was made on the Nottingham Show, due entirely to the outbreak of foot-and-mouth disease, which made it impossible to hold any of the livestock events. There were, however, certain compensations. The Queen consented to be the Society's Patron and 'for the first time in the history of the Society a Ladies' Committee was formed under the Chairmanship of Mrs Howard Lamin whose enthusiasm and energy inspired her Committee to a high pitch of efficiency, and relieved the Society's staff of such onerous work as the organisation of the Children's Nursery, reception of lady visitors accompanying distinguished visitors, arrangements for the Ambulance services, and last, but not least, the entire responsibility for the Royal Pavilion, including furniture and internal decoration. Mrs Lamin and her Ladies' Committee also arranged a magnificent Ball in the Astoria Ballroom, Nottingham, which was an outstanding success.'[14]

There was also, as there should have been at a Show held in the Robin Hood country, an Archery Competition. 'It is thought,' the Council reminded Members, 'that this is the first time such a competition had ever been seen in an Agricultural Showyard. Archers from all over the country took part. The preliminary heats were shot off in the morning of the last day, the winners competing for the finals in the Main Ring in the afternoon and evening. The prizes were presented to the winners by the Sheriff of Nottingham.'

But the Coronation Show, in 1953, held in Bath, produced a handsome profit of £14,000. This, it was felt, 'must surely go down in the history of the Society as a truly remarkable achievement,'[15] and the financial result was far beyond the Council's expectations. 'The immediate post-war Shows,' it was admitted, '—Cheltenham, Cardiff and Bristol—all showed greater profits but it must be remembered that these were held immediately after the War at a time when the general public, after seven years of austerity, were in a mood to enjoy themselves at such events as an Agricultural Show. Since then there has been a gradual falling off of attendances at all Shows.'

The weather was cold, but dry, and 100,000 people paid to attend the Show. A number of new features had been arranged to instruct and entertain them. The Secretary was always a great man for superlatives, as any successful publicist must be, and it was noted in the Annual Report that 'for the first time at any Agricultural Show a special shed was erected to house the Champion Animal of each Breed of Cattle. It was hoped that this would prove to be a popular innovation as it was thought that it would be of benefit to those visitors interested in Stock to be able to see the Champion Animals in one shed instead of having to wander through the lines to look for them. Unfortunately, the Exhibitors and Stockmen do not favour this new departure. Some of the animals became very restive and the Stockmen had to be with their Champion Animals throughout the day, and those who had more than one animal in the Show found it impossible to look after their charges efficiently.'

There was a Church of England Pavilion, organised by Col. Garton and the Somerset Guild of Craftsmen. 'The purpose of the Pavilion was twofold,' it was recorded, 'to provide a Chapel in which to offer daily, to those engaged in the Show, the ministrations of the Church and to demonstrate, by means of a Photographic Exhibition, the richness and glory of our heritage in the National Church.'

The Bath Federation of Townswomen's Guilds had a stand which consisted of 'My Lady's Boudoir' as it was in 1800, with period costumes and furnishings, contrasted with 'My Lady's Boudoir' in 1953. The Mayor of Bath organised a coach run from Bath to London to present Loyal Greetings to Her Majesty the Queen from Bath, Chippenham, Calne, Marlborough and other towns. The coach, with passengers in period costume, paraded in the Show ring before beginning its journey to London. In connection with the Coronation festivities the Society was asked if it would contribute a tableau in the Coronation Procession which took place in Bath on the evening of Coronation Day, and with the help of the Young Farmers' Clubs a tableau was staged depicting the first meeting of the Society in 1777. This won first prize.

The Annual General Meeting of the Society in 1952 was held in Abbey Church House, Bath, where the first meeting had taken place in 1777. At that meeting, Lord Bledisloe said he would like to present a plaque to commemorate that historic event. The offer was accepted, and the plaque was unveiled in Coronation Year on November 5, in the presence of Lord Hylton (the retiring President), Lord Eldon (the new President), the Mayors of Bath and Exeter, and a number of prominent Bath citizens.

Exeter in 1954, when the Queen Mother attended, yielded a good profit. Exeter had been fortunate, since the weather that summer was abominable and the Bath & West proved to have chosen the only completely fine week in the year. 'Sympathy,' said the Council, and no doubt it meant it, 'must be extended to other Societies, particularly to the Three Counties and the Royal Welsh, whose Shows were held under terrible conditions'.

For some reason, however, the Livestock Sections at Exeter as in previous years were not supported as the Society felt they should have been. Efforts to improve matters had not been successful. 'The Society,' commented the Council, 'has for many years, endeavoured to encourage breeders to exhibit at the Bath & West. As a further incentive the Society decided some years ago to permit Stock entered in the Open Classes to leave the Showyard on the Friday evening, reserving the last day for Local Stock, and prize money in many Classes was also raised without any increase in Entry Fees. The interests of the Stockmen has always received the consideration of your Council. Excellent sleeping accommodation and plenty of hot water are provided along with a Marquee for food and refreshment.

'Despite all these efforts the total entry of Horses, Cattle, Sheep and Pigs was disappointing. It may well be that shortage of labour and the heavy cost of having their stockmen away from home for nearly a week make it difficult for Exhibitors to enter their Stock at Shows outside their own districts.'[16]

But this year, for the first time since the war, classes for cheese, butter and cream were included in the Prize Schedule, and 'the entries in each section were so numerous that at Launceston next year, the Marquee is being increased from 20 ft. to 40 ft. deep.'

However water, not space, turned out to be the main problem at Launceston.

When the site was originally inspected, it was obvious that the existing water supply would be quite inadequate. To meet this difficulty, the Society commissioned two bore holes, a measure, it pointed out, 'which might appear more becoming to the Royal Agricultural Society of Kenya'. The wells proved entirely satisfactory.

The Ring events depended a good deal on the internal combustion engine. The Royal Navy had undertaken to provide a helicopter rescue act, but 'the pilot succumbed to temptation and amused the crowd by taking on some of the show jumps. His machine generated so much wind, however, that the obstacles were often blown to the ground and our announcer awarded four faults.'[17] The Royal Corps of Signals gave motorcycle trick riding displays. Rain had made the ground very treacherous, 'but the Motor Cyclists displayed admirable skill and carried out their performances with the utmost resolution'.

In 1957, after the Swindon Show, Sir Charles W. Miles, Bt., resigned as Honorary Show Director, a post he had held for many years. The decision, Sir Charles said, had been forced on him by increasing deafness. Col. W. Q. Roberts was appointed to succeed him. Sir Charles' resignation came in a year when the Council had set up a special Sub-Committee to look into the finances of the Society and to make suggestions as to how economies might be made without affecting the work of the Society and its Annual Show. It was pointed out that increases in the cost of wages and materials had caused the Show expenses to rise by 58 per cent since 1937. These 'alarming increases', Members were told,[18] had 'made it necessary to explore the advisability of taking over land as a permanent site after the Show of 1960. The Sub-Committee have inspected a site forming part of Ashton Court Estate, Bristol, and have approached the Agents to the Estate to ascertain whether their clients would agree to sell the land to the Society, and on what terms. Negotiations are still taking place. This does not in any way preclude considering other sites within the area of the Society's activities.'

At Swindon in 1957 it was a case of third time lucky. The Society's first visit, in 1906, had been marred by a tram accident, in which several people had been killed and many injured, and in 1929 the attendance had been seriously reduced by outbreaks of smallpox in the district.

One of the outstanding features of the Swindon Show was the 'Farming Yesterday, Today and Tomorrow' exhibition, organised by the Ministry of Agriculture. 'The cost, which was mutually borne by the participants, was,' the Council admitted, 'so high that it is unlikely that the exhibition will be reproduced at the Bath & West or any other Agricultural Show in the country.'[19]

'Farming Yesterday' showed some of the farming methods used in the eighteenth and nineteenth centuries; the 'Farming Today' section was concerned with agricultural education and with the technical advice available to farmers, and 'Farming Tomorrow' indicated some of the possibilities for the future which might result from research. The demonstrations included flailing, a ploughing match between a team of oxen and a radio-controlled tractor, and Machinery of the Future.

In a special part of the Trade Section, R. A. Lister and Co. had organised a heavy machinery Trade Show, 'the like of which has never before been seen at an Agricultural Show. Such machines as portable pumping sets, portable air-compressors, concrete-mixers, winches, portable stone-crushers, cement grouting machine, grain-conveying plant, mobile conveyor, mobile crane, and many others were included in this highly interesting exhibition. Messrs. R. A. Lister & Co. engaged special trains from London and elsewhere to bring manufacturers of these machines and many of their prospective clients to the Show. The Society had the privilege of entertaining, on the third day, Lord Mills, the Minister of Power, who evinced keen interest in this part of the Trade Section.'

The Bath & West had a spectacular machinery demonstration of its own at each Show now, in the form of its electricity plant first used at Exeter in 1954. The building to contain it has a prefabricated sectional steel frame. To this was fixed an outside skin of hardwood panels and an inside skin of hardboard. It was painted outside in the Society's colours of cream, green and red. The generating plant was described in the most glowing terms in the Show Catalogue. 'Three huge engines, each 600 h.p., generate sufficient electricity—light and power—to supply a small town, and from the waste heat from these engines 3,000/4,000 gallons of hot water per hour are circulated through rubber pipes to all parts of the Showyard. These engines, which the general public can see in operation, are housed in a specially constructed building of steel, the inside of which is very attractively designed and furnished.' Thousands of visitors to the Showyard went to look inside the generating station, and paid 6d a head for the privilege. There was a viewing gallery extending along the full length of the power house and display panels showed details of the various uses of diesel and electrical plant in industry. There was also a working model of the waste heat recovery system.

Discussion on a permanent show-site went on throughout 1958 and 1959. In 1959 the President assured Members that 'only Bristol and Bath had been in Council's mind for a permanent site. One or two other places had been suggested, but for various reasons they were found to be impracticable.' A permanent showground had to be found, 'because it was not only a question of expense, but also the advisability of improving the amenities of the Show. Visitors would not put up much longer with the rather primitive services which they were bound to get when a show was on a site for only a year.'[20]

The following year the Show was in Bath, where the attendance was the lowest since the war, causing a deficit of more than £13,000 on the year's working. At the Annual General Meeting, the Chairman was asked if there was any further news concerning a permanent site, and gave the answer that 'certain negotiations were proceeding and he thought as a result of this that there were hopes of a site being found in or near Bath'.[21]

Another of the peripatetic shows was held at Ashton Court in Bristol in 1961. The weather was perfect, but the attendance was only 75,000, compared with 122,000 in 1949. 'Various reasons,' said the Council, 'have been given for the decline—haymaking, Derby Week and counter-attractions being those most frequently suggested. It must be for consideration, however, that the public have now quenched their immediate post-war thirst for entertainment and are weary of the stereotyped Agricultural Show.'[22]

One of the most disappointed, and possibly baffled people must have been the Secretary, J. G. Yardley, who retired on June 30, 1961, immediately after the disastrous Bristol Show. Always and essentially a showman, he was a man of original ideas and great drive and much of the credit for the post-war boom, when Bath & West attendances created records, must go to him. He did not, however, in any sense retire, since he continued to be the Secretary of the Royal Smithfield Club.[23]

The post of Secretary had been advertised in the previous year, and Lord Darling, who had been Assistant Secretary since 1954 was appointed from eighty applicants. At the same time John Davis, who had first come to the Society in 1936, was appointed Assistant Secretary.

What proved to be almost the last of the round-the-region Shows were both held at Taunton, in 1962 and 1963. Neither was, or could have been, a spectacular success, but there were several interesting pointers to the future. Chief among these was the sponsorship of show-jumping events by outside bodies, in order to increase the scale of the prize-money and consequently the willingness of leading horsemen to compete. The TWW Stakes, the Commercial Union Stakes and the Guinness Time Stakes offered a total of £1,000 in prizes.

14,000 fewer people came to the Show in 1963, compared with the previous year, and once again the Council spent many hours scratching its head and wondering why. This time it decided to blame the failure on television, 'which brings a wide range of amusement which can be obtained with the minimum effort'.[24] But, with plans for the 1964 Show at Swindon being made, it was announced that 200 acres of flat land near Shepton Mallet was to be the Society's permanent home and that the 1965 Show would take place there.

The Swindon Show was, in the opinion of most of those responsible for it, best forgotten. The weather was appalling, the ground was a sea of mud and the Society finished up with a £26,000 deficit. And so, said the Secretary, 'let us leave the coverts which have been drawn in the past from nearly Land's End to South Wales, up to Birmingham, across to Wimbledon, down through Kent and return to the heart of the West Country'.[25] While the contractors and the visitors at

56. The Secretary, Lord Darling, on the newly acquired permanent show-site.

Swindon had been coping with Flanders-like mud, work had been going on relentlessly at the Shepton Mallet site. Hedges had been removed, ditches piped, roads and entrances made. In June 1965, the Society held its first Show there.

The 1965 Show Catalogue outlined what had been done in the way of preparatory work during the previous twelve months, pointing out that it had been the Council's policy to give the work to local contractors wherever possible. Two miles of hedges had been ripped out, two and a half miles of land draining laid, nearly a mile of new boundary fence erected, one and a half miles of water-mains installed and 15,000 square yards of macadam laid. This, however, was only a beginning and each year visitors would be able to see the improvements that had been made during the twelve months since they were last there.

The very fat Catalogue—516 pages—gives a clue to the size of the Show in this watershed year. The recipe was, for the most part, 'roughly as before, but more of it'. The Ring Events, however, included a number of novel items. Among these was a competition, or perhaps one should say part of a competition, between two teams of Ford tractor demonstrators, the Ford Demons and the Blue Angels. 'The competition,' announced the Catalogue, 'which will be held over a series of contests staged at the more important agricultural shows throughout the British Isles during the summer months, consists of a variety of hair-raising stunts, which will not only test the drivers' skill but will also demonstrate some of the amazing qualities of the new range of Ford tractors.

'Making its public debut during the Show will be a radio-controlled version of the Ford Super Dexta 3000, fitted with the revolutionary Select-O-Speed change on the go-transmission. This driverless, clutchless tractor will fascinate spectators with its almost magical demonstration.

'As audience reaction will be taken into account by the judges, it is hoped that audiences will enthusiastically support their favourites and join in the spirit of this sporting occasion.'

The displays by the RAF Abingdon Sport Parachuting Club was equally impressive. 'The displays,' it was promised, 'will take the following pattern: the aircraft (Rapide) will fly into wind across the dropping zone at a height of 2,000 feet above the ground and a wind-drift indicator will be released directly over the landing target on the ground. The aircraft will then continue its climb to 9,000 feet. At this height the aircraft will again run into wind across the dropping zone and four to six parachutists will jump out. After leaving the aircraft they will demonstrate their aerobatic manoeuvrability during their forty-five seconds of free fall. Two of them will then attempt to link up and continue their fall holding hands. Each jumper will be wearing a smoke generator strapped to his leg so that the manoeuvring is easily seen by the spectators. At 2,000 feet parachutes will be opened and the remainder of the display will demonstrate the techniques of steering into a target (a cross) on the dropping zone in the Main Ring.'

The internationally famous sheepdog demonstrator, Mr W. John Evans, was also there, with six dogs. 'Supreme International Champion in 1953 and 1958, Mr Evans has also won over two hundred Open Championship Trials, including winner four times of the Royal Welsh Stakes, Daily Express International Trials at Hyde Park, London, and twice winner of the Gold Cup at White City, London. Apart from the normal demonstrations, Mr Evans will receive commands over the loudspeaker from the commentator. These he will pass on to the dogs so that the public may get a closer understanding of the ways in which he controls his dogs.'

With the move to Shepton Mallet and the very substantial investment involved—by the end of 1966 over £100,000 had been spent on buying and developing the Show site—the Society had been compelled to face up to the fact that it was running a large business, and that the old form of organisation was not adequate to meet the new demands. At the Annual General Meeting in 1965 an Executive Board, later known as the Board of Management, was set up to manage the overall affairs of the Society. This was to consist of not more than twelve people, elected by the Council from its own members. The Chairman from the beginning was Sir Gerald Beadle, and at the Annual General Meeting in 1966 he explained the policy the Board intended to follow.

57. The Showground from the air, 1969.

It was not, he emphasised, set up to run the Bath & West Show. That, he said, 'continues to be run, as it has from time immemorial, by the Stewards and their specialised Committees, under the co-ordinating authority of the Honorary Show Director'. The function of the Board was quite specific. 'The Board,' Sir Gerald explained, 'is the custodian of the property, and one of the Board's duties is to manage and develop the property in the interests of the Show. But development needs capital. A year ago we launched an appeal for a Capital Fund of £150,000. I am glad to be able to tell you that, in cash and promises, we have received about £90,000 so far—getting on for two-thirds of our original target.'

The capital was to be used to increase the Society's profitability. 'Our survival in the long run,' Sir Gerald went on, 'depends on our income being not less than our expenditure—preferably more. Capital projects must be aimed at one of two objectives—either reduced expenditure or increased income. Every capital project must have one or another of these two objects in view. In the recent past our balance of income and expenditure has not been satisfactory. This is what led us to go permanent and to seek a capital fund by gift instead of by loan. We have done both. Now we must make sure that the capital fund is applied in such a way as to bring our balance of income and expenditure into a satisfactory state. That is the problem before your Board, and that is one of the reasons why you need some first class business brains on the Board.'

'It would,' he pointed out, 'be bad business and bad public policy to occupy these magnificent two hundred acres permanently with nothing in mind but a four-day annual Show. We can't farm it in the proper sense because we need it for the Show at the height of the farming year. To use it merely as a sheep ranch would be wasteful. We must devise and encourage suitable and profitable uses for the Showground all the year round.'

The Secretary added his own views on this. 'It would,' he was sure, 'be a less hair-raising and harrassing life for many of us if there was no Show but simply a self-supporting succession of Agricultural events. As it happens, the Society has grown into a way of life which produces a situation whereby its only hope of profit is also its greatest hazard—namely, the Show. A permanent Show site should lessen the element of risk. The first few years must, however, be a period of development during which it will not be possible to cash in on the benefits of permanency to any great extent.

'A simple example of this would be permanent roads. In the past, sleepers were bought and sold annually, sometimes at a profit. Now some £10,000 is locked up in roads which, bearing in mind that the Society is a registered charity might have been earning £700 a year net. That £10,000 might have been spent economically on constructions which cost the Society considerable annual hire charges. Absence of roads on a permanent site would not, however, be tolerated by Members, the Public, nor, after 1964 experiences, by the Council.

'Somehow, "out of Show" profitable uses for the site must be found.'

In 1967 it was possible to announce a number of outside-the-Show developments. On May 24 a commercial poultry unit was installed on the site, housing 5,000 laying birds and demonstrating 'the latest developments in poultry housing, husbandry, and cage manufacture'. This unit was to be open all the year round. Another important project was the arrangement to allow Palmer, Snell and Co., the auctioneers, to hold sales of second-hand agricultural machinery on the site.'

These enterprises produced much needed income, and there was no opposition to them. Another money-raising effort in 1967 proved more controversial, however. The story is to be found in the 1967 Annual Report. In 1800, 'friends of the artist Thomas Robins the younger asked the Society one hundred guineas for the hundred drawings of flowers and insects. Their idea was to obtain relief for the artist, who had fallen on hard times. The Council, wishing to help, offered twenty guineas, giving Robins the option to buy back within two years. The poor fellow soon died and the Society became outright owners of the drawings which, bound in two volumes, remained in the offices until 1966.

'It was observed that five of the drawings had been removed over the years, but still no great importance was attached to them. In fact, a local valuation in 1966 placed £750 on the two volumes. Nothing more than a hunch caused discussions with Messrs. Sothebys and eventual arrival of the pictures at their premises. Upon being deposited, they were immediately insured for £8,000 [the figure was actually £20,000], and for the first time there was realisation that

Dame Fortune might be smiling. To sell or not to sell became the debate. The Council agreed, against the recommendation (very understandably) of the Art Committee, but, by a huge majority, to sell and to leave detailed arrangements to the Board of Management. Eventually half the drawings were auctioned at Sothebys in July 1967. The gross expectation was around £4,000. The Chairman of the Board of Management and the Steward of Art were flies on the wall at the sale. Imagine their amazement when the first single picture went for £400, the second pair for £800, and so on. The consignment (half the total possession) fetched £13,893 net, at an average of £348 per picture. In October six more were sold by Messrs. Sotheby in Toronto for a net average of £361 each.'

The Council's argument for selling the drawings was twofold, that it would be much better to have them where they could be seen and enjoyed, rather than buried in the Society's offices and that the money obtained from selling them could be put to good use. Both arguments were probably true, but the real bone of contention was the splitting up of a homogeneous collection of paintings, unique in its subject matter, into a number of separate wall-decorations.

Most of the proceeds of the sale were devoted to developing the Show site. Between the 1967 and 1968 Shows a number of major works were carried out. A sewerage system was installed. Flush lavatories were installed. More hard roads were laid. Three kitchens were built, one for herdsmen and two for Members. Herdsmen's cubicles were constructed. Permanent carcass and cheese pavilions replaced the previous tented accommodation and money was somehow found for an Art Pavilion. Two handymen were added to the permanent staff, and they spent much of their time making minor equipment which would otherwise have had to be hired.

A further £56,000 was spent on the site during 1968–69, on new roads, car-parks and cattle-sheds.[26] At this point, the Chairman of the Board of Management warned Members[27] that money was running out and future capital development would have to be financed from surpluses on the revenue account. The Society, in other words, had got to pay its way. Before 1967 this would have seemed unlikely, but in that year there was a surplus of £907 and in 1968 of £2,370. 'These,' as Sir Gerald admitted, 'were not very impressive figures in themselves, but they were an indication that the Society was moving in the right direction.'

In 1970 the Society embarked on its most celebrated and most criticised out-of-season activity so far, the Festival of the Blues. Tens of thousands of young people converged on the Showground from all parts of Britain, and from many other countries as well, and the appearance of this normally peaceful part of Somerset had to be seen and experienced to be believed. Hundreds slept in tents and thousands more on the open fields. The sanitary facilities were taxed up to and beyond the limit, feeding the multitude became a serious problem and the

58. Display at the Show, 1969.

police spent a good deal of time waiting for trouble which did not, in fact, materialise, although the appearance of many of the visitors gave rise to understandable misgivings among local people who were not used to having such exotic beings in their midst.

The Secretary's reference to the event, in the Annual Report, could hardly be beaten for tact. 'The question is continually being asked,' he said, 'as to whether another Festival will be held. The short answer is that no application to do so has been received. It seems highly probable, however, that promoters will be forced to re-think their approach to such events because, apart from the many administrative problems involved, it has apparently proved impossible to ensure the collection of entrance money from a large percentage of the attendance.'

For the other major out-of-season function, the *Farmer's Weekly* National Pig Day, the Society was rather better prepared. The new Cattle buildings provided all the accommodation required and 6,000 people attended. By now, however, it was being realised that the Society was not charging enough for its facilities. But what, the Secretary asked, was a fair price to charge? Somehow, one's natural wish to be helpful and to encourage as many organisations as possible to use the Showground during its fallow eleven months of the year had to be brought into line with the need to make money. And this balance could only be learnt by trial and error.

When the Board of Management was set up in 1966, it was emphasised that the Board had no responsibility for the Show. Five years later this was felt to have been a mistake. There was an efficient system for dealing with general business and a rather slow-moving and inefficient committee-system for supervising the Show and for reporting through the Finance and General Purposes Committee to the Council, with all the delays made inevitable by such a long management chain. In 1971, Council decided to elect an Executive Committee, to take over all the responsibilities of the Board and to deal with Show matters as well. This provided an integrated pattern of planning and control, because, as Sir Gerald Beadle pointed out at the 1972 Annual General Meeting, 'if the scope, size and standard of the Show is to be maintained, it must be subsidised from profitable out-of-Show activities'. These soon included caravan rallies, clay pigeon shoots, dog shows, a Spring Jumping Festival, a Women's Institute Jamboree, Pony Club camps and numerous Bath & West Riding Club events, and Steam Fairs.

'In view of the fact that the Society's (and some other) governing bodies are

59. Moving day at Pierrepont Street, May 1974.

60. New headquarters at Shepton Mallet.

often thought to consist of old or retired gentlemen,' reflected the Secretary, 'it may be of interest to record the professions of the first nine Elected Members to have been voted in. These are: four Farmers, two Land Agents, two Brewers, and a Solicitor.'[28] Old or retired gentlemen or no, the Executive Committee seems to have fully justified its existence. Under its guidance the bank loan, originally taken up to acquire the site, was steadily reduced, and membership rose each year, despite the overdue decision to raise the annual subscription.

Taking stock in 1972 of progress since the move to Shepton Mallet, the Secretary invited Members to take an objective look at the facts. 'Before the days of permanent sites,' he pointed out, 'Societies had no capital sunk in fixed assets, but hazarded themselves upon being able to meet vast contractors' bills.'[29] They should ask themselves four questions:

What would the contractors' bill for a peripatetic Show be now?
What would the rent be for temporary Show sites now—if we could get them?
Would Members, Exhibitors and Visitors accept the low standard of facilities inherent in moving Shows?
Can we now, given a bad year, make such a loss as we did at Swindon in 1964 (£26,000)?'

At the same time, the Secretary reported that new administrative offices were to be built at the Showground.[30] The move from Bath was to take place after the 1974 Show. Pierrepont Street, he reminded Members, had been the working home for thirty-seven years of the Assistant Secretary, John Davis, who had served under three Secretaries and who knew more, publishable and un-publishable, about the Society than anyone else alive.[31]

By 1973 it was possible to say with due caution, 'Things have come right for a couple of years'. The Society's profit in 1972 was £17,468 and in 1973 £16,798.[32] The recipe for the new-style Show—'a touch of the Hurdy Gurdy combined with an agricultural demonstration such as no one else attempts'—appears to be working, and although credit for the phrasing must go to the present Secretary,[33] Lord Darling, the formula itself would have surely met with the complete approval of the two other great showman-secretaries, Plow-man and Yardley. The Hurdy Gurdy will, of course, change its shape from decade to decade and the demonstration element, too, will move with the times. Twenty or thirty years from now, perhaps, the dairy cow will have had its day, with green crops being converted directly into milk, protein will be coming from soya-beans and petroleum and vast intensive pig-fattening units and broiler-chicken houses will be things of the past. When these developments occur, the Showground at Shepton Mallet will have a different look and the Hurdy Gurdy will be playing different tunes. But, come what may, these 200 flat acres are likely to prove a very sound investment.

Aims, Rules and Orders of the Society, 1777

'The principal object of this society's attention will be,

To excite by premiums a spirit of emulation and improvement in such parts of husbandry as seem most to require it:

To endeavour to increase the annual produce of corn, by bringing into cultivation, in the least expensive and most effectual manner, such lands as are at present barren or badly cultivated, particularly by draining and manuring; and by the introduction of various sorts of vegetable food for cattle:

To promote the knowledge of agriculture by encouraging and directing regular experiments on those subjects which are of the most importance to it, by distributing rewards to such persons as shall raise the largest and best crops both of natural and artificial grasses, and the several species of grain, on any given quantity of ground:

To encourage planting on waste lands, raising of quick-hedges, cultivating turnips, Scotch cabbages, &c. &c.

To promote all improvements in the various implements belonging to the farmer, and introduce such *new* ones as the experience of other counties has proved more valuable than those generally in use:

This society's attention will also be directed to all improvements of the machines used in our different manufactories, as well as the manufactures themselves; and to encourage ingenuity, diligence, and honesty, in servants and labourers:

And, to sum up the whole, every thing that is conducive to the prosperity of the counties of Somerset, Wilts, Glocester, and Dorset, and the good of the community at large, will be diligently attended to by this society.

N.B. The society's meetings will in future be held the second Tuesday in every month: And all letters relative to the various objects of its attention, directed to the secretary, will receive due notice and attention.

Rules and orders

I.

That a Monthly Meeting, consisting of not less than Nine Members, shall be held on the second Tuesday in every Month, for transacting the usual Business of the Society; but that no *New Laws* shall be made, nor the following ones altered, but at a General Meeting, to be held Annually on the second Tuesday in December; which Meeting shall not consist of less than Twenty Members.

II.

That the President, or, in his Absence, one of the Vice-Presidents, shall preside at, and regulate the Debates of all future Meetings; and that the Vice-Presidents shall take the Chair by Rotation.

III.

That after the Expiration of the Year 1777, all Committees shall be annually chosen at the General Meeting in December; and the said Committees shall be empowered to adjourn from Time to Time, as they may see Occasion: And that on any Vacancy, or Vacancies (by Death, Removal, or Resignation) being declared to the Secretary, he shall make Report thereof to the next Monthly Meeting, which shall fill up such Vacancy, or Vacancies, by appointing a new Member, or Members, to that Office.

IV.

That each Committee, when met, shall chuse a Chairman, skilled in the particular Business to which it is appointed; and when fitting, to enter Minutes of their Proceedings in a Book for that Purpose. That all Reports to the Society be made in Writing, and signed by the Chairman; and that the Secretary shall enter those Reports in the Society's Journal.

V.

That an Annual Subscription of any Sum, not less than One Guinea, shall entitle a Person to be a Member: And that the Names of all Persons who give Benefactions, not less than Half-a-Guinea, shall be published with the List of Members. That a Benefaction, not less than Twelve Guineas, shall entitle any Person to be a Member for Life.

VI.

That, for the future, any Person desirous of becoming a Member shall give or send to the Secretary his Name, Place of Abode, and the Sum he intends to subscribe: His Name shall then be entered on the List of Members, and his Subscription shall bear Date from the preceding Quarter-Day, and become due again that Day Twelvemonth.

VII.

That a List of such Premiums as the Society may think fit to offer shall be printed and published on or before the first of January in every year, which

Premiums shall be classed under the several Heads proposed to be encouraged by this Institution.

VIII.

That no Premium shall be offered until it has been first proposed to and approved by a Committee, and agreed to by the Annual Meeting. And no Premium or Bounty shall be given to any Candidate, unless the Society, at the Annual Meeting, shall be satisfied that such Candidate deserves it.

IX.

No Member of the Society, who is a Candidate for any Premium or Bounty, shall sit in any Committee to which such Matters may be referred, or be present while the Subject is under Consideration; nor shall such Candidate be present in the Meetings of the Society when the Matter is before them, whether in Debate, or for Determination, unless when called in to answer such Questions as may be put to him.

X.

That all Claims for Premiums or Bounties shall be made at least two Months before the Annual Meeting in December: And that such Claims must be given in to the Secretary in Writing, and by him presented to the next Meeting of the Committee to which they relate.

XI.

In order that all Rewards may be distributed with the utmost Impartiality and Justice, the Society shall, when they think it necessary, desire the Assistance of such Gentlemen, Manufacturers, Artists, or others (though not Members) as shall be deemed best able to judge of and discover the Value of any Invention or Improvement for which a Premium is claimed.

XII.

That Premiums shall be both honorary and pecuniary; but that no Premium or Bounty shall be given by this Society to any Person who shall have obtained a Premium or Bounty for the same Invention or Improvement from any other Society.

XIII.

That as the principal Design of this Institution is, by exciting a Spirit of Industry and Ingenuity, to promote the Public Good, the Premiums first offered shall be more immediately directed to Improvements in Agriculture, Planting, and such Manufactures as are best adapted to these Counties.

XIV.

That until a sufficient Fund be raised for offering pecuniary Premiums, the Society shall give *Honorary Rewards* for such Specimens of Ingenuity as they may be favoured with; and that, for this Purpose, Silver Medals be struck, expressive of the Nature and Design of this Institution.

XV.

That some Premiums be offered for the Encouragement of Industry and good Behaviour among Servants in Husbandry, and Labourers.

XVI.

That the Premiums offered in any one Year shall not exceed two-thirds of the Fund in Hand, at the Time the said Premiums are offered by the Society.

XVII.

That at the Annual Meeting, to be held on the Second Tuesday in December next, the Society's Cash shall be accounted for by the Secretary, and deposited as the Meeting shall then determine.

XVIII.

That in order to encourage the Study, as well as the Practice of Agriculture, &c. &c. Honorary Premiums shall be offered for the best-written and most useful original Essay on any of the Subjects to which the Views of this Society may be extended, that may be sent to their Monthly Meetings; the Society to give out the Subjects in their annual List of Premiums: And that such Essays as shall be approved of at the Annual Meeting be printed and published at the Expence of the Society. Every Member to have one Copy, and the Rest of the Impression to be sold, and the Profits applied to the Society's Use; unless the Author shall think proper to print the same at his own Expence; in which Case he shall send the Secretary as many Copies as there are Members, to be distributed amongst them.

XIX.

That the Authors of such Essays shall send them sealed to the Secretary, without a Name, but with some Mark, corresponding with another Mark on the Outside of an inclosed sealed up Paper, in which their Names are written: That such of the Essays as are rejected shall be left in the Secretary's Hand, together with the corresponding Papers not opened, and if they are not called for, shall be destroyed at the succeeding Annual Meeting.

XX.

A Candidate for a Premium, or a Person applying for a Bounty, being detected in any Attempt to impose on the Society, shall not only forfeit such Premium, or Bounty, but be declared incapable of obtaining any for the future.

XXI.

That the Secretary shall procure all such Books and Stationary Ware as are needful for the Society's Use, and keep fair Accounts of all Monies received and disbursed by him: The said Accounts to be settled and balanced at each Monthly Meeting in the Society's Cash-book, where a Committee of Accounts shall be appointed to audit them. He shall also perform the necessary Business of his Office with Diligence and Integrity, viz. Attend all Meetings and Committees of the Society;—make all Minutes and Resolutions, and enter them fairly in the Journal or Committee Books;—read all Letters and other Papers sent to the Society, and prepare such Answers thereto as the Society shall direct, and record regularly in the Book of Correspondence such as are worthy of Preservation;—sign all Publications, Notices, and Receipts, and prepare an Abstract or Annual Register of the Transactions of the Society, with particular Accounts of the Improvements made in the various Articles to which its Views are extended, together with such Informations received as appear to be the most useful and important, and the Resolutions of the Society thereon;—attend to the Printing of whatever the Society may direct to be printed, and correct the Press. He shall also collect Subscriptions, and visit Manufactories, or apply for particular information respecting them when required by the Society so to do; and as much as possible make himself acquainted with the various Arts, &c. &c. to which the Views of this Society shall be directed: He shall also regularly enter the Minutes, Proceedings, and Resolutions of each Meeting, for the Inspection of the next; and that, in Consideration of his Trouble, and the close Attention he must give to this Business, he shall be allowed an annual Salary.

XXII.

That on any Emergency the Secretary, with the Concurrence of five Members, signified in Writing, and signed with their Names, may call an Extra General Meeting, by Advertisement in the public Papers of the respective Counties.

XXIII.

All Letters relative to the Business of the Society, being laid by the Secretary before the Committee of Correspondence, that Committee shall be at Liberty from Time to Time to refer such Letters as they think proper to the other respective Committees, without waiting to report them to a Meeting of the Society; *unless*, such Letters relate to the granting of any new Premium or Bounty.

XXIV.

All the Books, Papers, and Correspondence of the Society shall remain under the Care of the Secretary, to be inspected by the Members at any reasonable Time.

XXV.

All Models of Machines and Implements which obtain Premiums or Bounties shall be the Property of the Society, and be kept for the Inspection of Farmers and Manufacturers.

XXVI.

In Case any Person shall be inclinable to leave a Sum of Money to this Society, by Will, the following Form is offered for that Purpose:

Item, I give and bequeath to A.B. and C.D. the Sum of   Pounds, upon Condition, and to the Intent that they pay the same to the Treasurer or Secretary, for the Time being, of a Society instituted at Bath 1777, who call themselves 'The Society for the Encouragement of Agriculture, Arts, Manufactures and Commerce'; which such Sum of   Pounds I will and desire may be paid out of my personal Estate, and applied towards carrying on the laudable Designs of the said Society.

XXVII.

Form of a Letter to Gentlemen whose Subscriptions are in Arrears

SIR,

I am directed to inform you, that your annual Subscription of   has been in Arrear since the   Day of   : And as it is of Consequence for the Society to know what Sums of Money they can bestow in Premiums, you are respectfully requested to order the Payment of it to the Secretary.

ERRATUM

The Secretary of the Society from 1938–1961 was
J. G. Yardley

Secretaries and Presidents of the Society

Secretaries

1777–1787	Edmund Rack
1787–1800	William Matthews
1800–1805	Nehemiah Bartley
1805–1818	Robert Ricards
1818–1849	Benjamin Leigh Lye
1849–1865[1]	Henry St J. Maule
1865–1882	Josiah Goodwin
1882–1919	Thomas Plowman
1919–1938	F. H. Storr
1961–1974	Lord Darling (Chief Executive since 1974)
1974–	John Davis

Presidents

1777–1780	Earl of Ilchester
1780–1798	Marquis of Ailesbury
1798–1800	Lord Somerville
1800–1802	1st Duke of Bedford
1802–1805	2nd Duke of Bedford
1805–1817	Sir Benjamin Hobhouse
1817–1847	Marquis of Lansdowne
1847–1852	Lord Portman
1853[2]–1854	Lord Portman
1854–1855	Earl Fortescue
1855–1856	C. A. Moody, MP
1856–1857	Lord Courtenay
1857–1858	Lord Courtenay
1858–1859	John Sillifant
1859–1860	Lord Rivers
1860–1861	James Butler, MP
1861–1862	Thomas Dyke Acland
1862–1863	Marquis of Bath

1863–1864	Earl Fortescue
1864–1865	Lord Taunton
1865–1866	Earl of Portsmouth
1866–1867	John Tremayne
1867–1868	Sir J. T. B. Duckworth
1868–1869	Earl of Carnarvon
1869–1870	Sir Stafford H. Northcote, MP
1870–1871	Earl of Cork and Orrery
1871–1872	Duke of Marlborough
1872–1873	Earl of Mount Edgcumbe
1873–1874	Sir Massey Lopes, MP
1874–1875	Richard Benyon, MP
1875–1876	Earl of Ducie
1876–1877	Marquis of Lansdowne
1877–1878	Earl of Jersey
1878–1879	Earl of Morley
1879–1880	Earl of Coventry
1880–1881	Marquis of Abergavenny
1881–1882	Lord Tredegar
1882–1883	Lord Brooke
1883–1884	Viscount Holmesdale
1884–1885	Viscount Hampden
1885–1886	Lord Carlingford
1886–1887	Earl of Ilchester
1887–1888	Lord Tredegar
1888–1889	Lord Clinton
1889–1890	Earl of Darnley
1890–1891	Earl Temple
1891–1892	Sir T. W. Llewelyn
1892–1893	Lord Fitzhardinge
1893–1894	Earl of Onslow
1894–1895	Viscount Portman
1895–1896	Earl of Clarendon
1896–1897	Lord Montagu of Beaulieu
1897–1898	Lord Windsor
1898–1899	Lord Clinton
1899–1900	Marquis of Bath
1900–1901	Earl of Cork and Orrery
1901–1902	Earl of Morley
1902–1903	Duke of Beaufort
1903–1904	Lord Windsor

1904–1905	Duke of Portland
1905–1906	Earl of Radnor
1906–1907	H.R.H. The Prince of Wales
1907–1908	Lord Digby
1908–1909	Lord Clinton
1909–1910	Earl of Darnley
1910–1911	Marquis of Bute
1911–1912	Marquis of Bath
1912–1913	Viscount Falmouth
1913–1914	Sir J. T. D. Llewellyn
1914–1919	Earl of Coventry
1919–1920	Earl of Radnor
1920–1921	Lord Bledisloe
1921–1922	H.R.H. The Prince of Wales (Deputy President: Lord Clinton)
1922–1923	H.R.H. The Prince of Wales (Deputy President: Lord Blythswood)
1923–1924	Sir Dennis F. Boles
1924–1925	Col. F. S. W. Cornwallis
1925–1926	Earl of Clarendon
1926–1927	The Duke of York
1927–1928	Lt. Col. Lord Wynford
1928–1929	Maj.-Gen. T. C. P. Calley
1929–1930	Lord Mildmay
1930–1931	Lord Digby
1931–1932	Lord Digby
1932–1933	Sir George Roberts
1933–1934	Mrs G. Herbert Morrell
1934–1935	Viscount Portman
1935–1936	The Earl of Jersey
1936–1937	The Duke of Kent (Deputy President: the Duke of Somerset)
1937–1938	Lord St Levan
1938–1939	The Duke of Somerset
1939–1947	Lord Dulverton
1947–1948	The Marquis of Bute
1948–1949	The Duke of Beaufort
1949–1950	Lord Willoughby de Broke
1950–1951	Col. the Lord Digby
1951–1952	Lord Belper
1952–1953	Lord Hylton
1953–1954	The Earl of Eldon
1954–1955	Lt. Col. Sir Edward Bolitho

1955–1956	Col. Sir Godfrey Llewellyn
1956–1957	Lord Herbert
1957–1958	Lord Roborough
1958–1959	Col. H. C. C. Batten
1959–1960	Lord Hylton
1960–1961	Sir Reginald Verdon-Smith
1961–1962	The Bishop of Bath and Wells (the Rt. Rev. E. B. Henderson)
1962–1963	Sir Bernard Waley-Cohen
1963–1964	Earl of Radnor
1964–1965	Lt. Col. J. A. Garton
1965–1966	Col. Sir John Carew Pole
1966–1967	Col. J. W. Weld
1967–1968	Lt. Col. the Lord Wigram
1968–1969	Col. the Lord Margadale of Islay
1969–1970	Lord Ashburton
1970–1971	Col. C. T. Mitford-Slade
1971–1972	Brigadier the Lord Tryon
1972–1973	Lord Dulverton
1973–1974	Earl Waldegrave
1974–1975	Col. C. T. Mitford-Slade
1975–1976	Lord Digby
1976–1977	H.R.H. The Prince of Wales

Membership, Income and Expenditure at 10-yearly intervals, 1859 onwards[1]

1859   Members: 1,258
Subscriptions: £954 19s 6d
Barnstaple Show:
    Receipts: £2,922
    Expenditure: £2,311
    Subsidies from Barnstaple and Tiverton: £906
Journal expenditure: £395, including Editor's honorarium, £25

1869   Members: 994
Subscriptions: £1,346 3s 10d
Taunton Show:
    Receipts: £5,629
    Expenditure: £5,281
    Subsidy from Taunton: £900
Journal expenditure: £714, including Editor's honorarium, £175
Journal receipts: 6/–

1879   Members: 967
Subscriptions: £987
Exeter Show:
    Receipts: £6,310
    Expenditure: £7,974
    Subsidy from Exeter: £800
Journal expenditure: £449 4s 6d
Journal receipts: £12 10s 0d

1889   Members: 1,205
Subscriptions: £1,180
Rochester Show:
    Receipts: £7,826
    Expenditure: £7,365
    Subsidy from Rochester: £800
Journal expenditure: £559
Journal receipts: £7

1899 Members: 1,214
Subscriptions: £1,130
Bath Show:
 Receipts: £8,751
 Expenditure: £8,650
 Subsidy from Bath: £800
Journal expenditure: £448 12s 0d
Journal receipts: £37

1909 Members: 1,131
Subscriptions: £1,108
Rochester Show:
 Receipts: £9,691
 Expenditure: £8,662
 Subsidy from Rochester: £800
Journal expenditure: £407 18s 11d
Journal receipts: £48

1919 Members: 871
Subscriptions: £770
No Show
Journal expenditure: £495
Journal receipts: £49

1929 Members: 1,164
Subscriptions: £1,175
Swindon Show:
 Receipts: £11,925 9s 10d
 Expenditure: £12,812 0s 0d
 Subsidy from Swindon: £1,200
Journal expenditure: £384
Journal receipts: £43

1939 Members: 1,131
Subscriptions: £1,059
Bridgwater Show:
 Receipts: £17,260
 Expenditure: £13,961
 Subsidy from Bridgwater: £600
Journal expenditure: £252
Journal receipts: £17

1949   Members: 3,000
Subscriptions: £3,253
Bristol Show:
    Receipts: £62,607
    Expenditure: £36,848
    Subsidy from Bristol: £3000

1959   Members: 2,679
Subscriptions: £2,993
Yeovil Show:
    Receipts: £59,301
    Expenditure: not published
    Subsidy from Yeovil: £3,000

1969   Members:
Subscriptions: £10,362
Shepton Mallet Show:
    Receipts: £81,240
    Expenditure: £71,741

Places where the Annual Show was held

Attendance[1]

1852	Taunton	
1853	Plymouth	
1854	Bath	
1855	Tiverton	
1856	Yeovil	
1857	Newton Abbot	
1858	Cardiff	
1859	Barnstaple	
1860	Dorchester	23,000
1861	Truro	29,000
1862	Wells	15,000
1863	Exeter	35,000
1864	Bristol	88,000
1865	Hereford	52,000
1866	Salisbury	26,000
1867	Salisbury	24,000
1868	Falmouth	31,000
1869	Southampton	57,000
1870	Taunton	52,000
1871	Guildford	34,000
1872	Dorchester	34,000
1873	Plymouth	62,000
1874	Bristol	110,000
1875	Croydon	41,000
1876	Hereford	49,000
1877	Bath	76,000
1878	Oxford	39,000
1879	Exeter	55,000
1880	Worcester	46,000
1881	Tunbridge Wells	46,000
1882	Cardiff	63,000
1883	Bridgwater	48,000

1884	Maidstone	45,000
1885	Brighton	49,000
1886	Bristol	100,000
1887	Dorchester	39,000
1888	Newport (Monmouthshire)	53,000
1889	Exeter	53,000
1890	Rochester	52,000
1891	Bath	76,000
1892	Swansea	73,000
1893	Gloucester	55,000
1894	Guildford	38,000
1895	Taunton	43,000
1896	St Albans	34,000
1897	Southampton	42,000
1898	Cardiff	56,000
1899	Exeter	56,000
1900	Bath	49,000
1901	Croydon	41,000
1902	Plymouth	54,000
1903	Bristol	109,000
1904	Swansea	79,000
1905	Nottingham	55,000
1906	Swindon	50,000
1907	Newport (Monmouthshire)	54,000
1908	Dorchester	33,000
1909	Exeter	57,000
1910	Rochester	26,000
1911	Cardiff	59,000
1912	Bath	55,000
1913	Truro	58,000
1914	Swansea	86,000
1915	Worcester	36,000
1916	no Show	
1917	no Show	
1918	no Show	
1919	no Show	
1920	Salisbury	45,000
1921	Bristol	100,000
1922	Plymouth	58,000
1923	Swansea	115,000
1924	Taunton	46,000

1925	Maidstone	38,000
1926	Watford	23,000
1927	Bath	50,000
1928	Dorchester	45,000
1929	Swindon	30,000
1930	Torquay	51,000
1931	Bristol	60,000
1932	Yeovil	45,000
1933	Wimbledon	51,000
1934	Oxford	31,000
1935	Taunton	53,000
1936	Neath	66,000
1937	Trowbridge	30,000
1938	Plymouth	85,000
1939	Bridgwater	56,000
1940	no Show	
1941	no Show	
1942	no Show	
1943	no Show	
1944	no Show	
1945	no Show	
1946	no Show	
1947	Cheltenham	104,000
1948	Cardiff	162,000
1949	Bristol	122,000
1950	Birmingham	72,000
1951	Dorchester	93,000
1952	Nottingham	73,000
1953	Bath	100,000
1954	Exeter	110,000
1955	Launceston	76,000
1956	Cardiff	110,000
1957	Swindon	108,000
1958	Plymouth	106,000
1959	Yeovil	66,000
1960	Bath	65,000
1961	Bristol	75,000
1962	Taunton	75,000
1963	Taunton	61,000
1964	Swindon	51,000
1965	Shepton Mallet permanent show-ground	60,000

1966	64,000
1967	67,000
1968	68,000
1969	89,000
1970	75,000
1971	76,000
1972	87,000
1974	95,000
1975	110,000
	103,000

Notes

Chapter One: GETTING THE SOCIETY LAUNCHED

[1] *English Farming Past and Present*, ed. Hill, 1951, p. 148. Originally published 1812.

[2] 'The Leicestershire Farmer in the Seventeenth Century', *Agricultural History*, XXV (1951).

[3] 'Coke of Norfolk and the Agricultural Revolution', *Economic History Review*, 2nd series, VIII (1955–56).

[4] *Nottinghamshire in the Eighteenth Century*, 2nd ed. 1966. Chapter 7.

[5] 'Opposition to Parliamentary Enclosure in Eighteenth Century England', *Agricultural History*, XIX (1948).

[6] See especially his critical article, 'Agricultural Economic Growth in England, 1600–1750', *Journal of Economic History*, 1965, reprinted in *Agriculture and Economic Growth in England, 1650–1815*, ed. E. L. Jones, 1967, and his book, *Development of English Agriculture, 1815–73*, 1968.

[7] A. Fothergill, 'On the Application of Chemistry to Agriculture and Rural Economy,' *Letters and Papers of the Bath Society*, Vol. III, 1786.

[8] *Report to the Board of Agriculture: Oxfordshire*, 1809.

[9] *Letters and Papers on Agriculture, Planting, etc.* Selected from the Correspondence Book of the Society. 1780.

[10] On the early societies, see Lord Ernle: *English Farming Past and Present*, p. 209 and Kenneth Hudson: *Patriotism with Profit*, 1972, pp. 1–24.

[11] 1791. Vol. I, pp. 77–81.

[12] The Racks had been Quakers for a long time. In 1665 an earlier Edmund Rack was prosecuted at the Quarter Sessions in Norwich for 'unlawful Meeting under colour and pretence of exercise in Religion, contrary to the Liturgy of the Church of England'. On this, see a pamphlet (1666), *The Norffs President of Persecution (unto Banishment Against some of the Innocent People call'd Quakers*.

[13] *Poems on Several Subjects*, London: Richardson and Urquhart, 1775; *Essays, Letters and Poems*, Bath. Printed by R. Crutwell for the Author, 1781.

[14] *The Life and Character of William Penn, Esq., Original Proprietor of Pennsylvania*, included in *Caspipina's Letters*. Bath: R. Crutwell, 1777.

[15] *Mentor's Letters, addressed to Youth*, 3rd ed., revised and corrected. Bath: R. Crutwell, 1778.

[16] 'On the Origin and Progress of Agriculture', in *Georgical Essays*, ed. A. Hunter, York: 1803.

[17] Curtis became one of the Society's Vice-Presidents. He died in 1784, and a tribute to him, written by Rack, was published by Crutwell in that year, with the title, *A Respectful Tribute to the Memory of Thomas Curtis, Esq.*

[18] Given in full, together with the Rules and Orders, as Appendix One, below.

[19] Pigs with the falling-disease or falling-evil, i.e. epilepsy.

[20] The General Meeting of 1783 agreed a premium of 10 guineas to 'the most numerous *Friendly Societies* of handicraftsmen and labourers who shall in the year 1784 be established in any country town or parish where no such society has hitherto been instituted'.

[21] The more important of these items are: *Minute-Book*, 1778, p. 30: Report on specimens submitted by Robert Davis; *Letter-Book*, p. 101, Davis's letter on cultivation; *Letter-Book*, p. 114, Cultivation notes supplied by Thomas Beevor; *Letter-Book*, p. 145, letter from Samuel More, of the London Society, who refers to the best specimen that Society had received as coming from Sir Alexander Dick of Edinburgh, who received a gold medal for it; *Letters and Papers*, Vol. III, 1786, Dr Falconer's report, Dr Parry's experiments with 'patients of the Pauper Charity' and article by Dr Hope, of Edinburgh, saying that the Royal Infirmary there had not for several years used any rhubarb other than that grown in Scotland. Also, in the same volume, a letter from Clarke, Jacam and Clarke, druggists, of London, referring to specimens raised by the Duke of Athol, 'from *true* Siberian seed'.

[22] See *Minutes*, Vol. V, p. 39, for the story.

[23] 'Letters and Papers on Agriculture, Planting, etc. Selected from the Correspondence-Book of the Society, instituted at Bath, for the Encouragement of Agriculture, Arts, Manufactures and

Commerce, within the Counties of Somerset, Wilts, Glocester, and Dorset, and the City and County of Bristol. . . .

Bath: Printed by R. Cruttwell, by Order of the Society, and sold by C. Dilly, in the Poultry, London, and by the Booksellers of Bath, Bristol, Salisbury, Glocester, Sherborne, Exeter, etc.'

[24] Paring off and burning the rough moorland turf in order to improve it.

[25] *Letters and Papers*, Vol. II, 1783, pp. 262–263.

[26] 'On the Application of Chemistry to Agriculture and Rural Oeconomy', *Letters and Papers*, Vol. III, 1786.

[27] *Letters and Papers*, Vol. III, 1786, p. 275.

[26] 'To prevent or keep meat from putrifaction', *Letters and Papers*, Vol. II, 1783, pp. 297–331.

[29] *Letters and Papers*, Vol. I, 1780.

[30] At the General Meeting in December 1781, it was announced that many of the members were 'several years in arrears'.

[31] Severely damaged in an air-raid during the 1939–45 War and subsequently pulled down. The site is now occupied by part of the new buildings of the Bath Technical College.

[32] 28th February 1787.

[33] Collinson's *History of Somerset*, p. 79.

[34] 'Inequalities of the Penal Laws', in *Essays, Letters and Poems*, 1781.

Chapter Two: 1787–1805: THE NEXT TWO SECRETARIES

[1] *The Life and Character of Thomas Letchworth, a minister . . . among the people called Quakers*, Bath: Cruttwell, 1786.

The Miscellaneous Companions, 3 vols., Bath: Cruttwell, 1786.

    Vol. I   being a short Tour of observation and sentiment through a part of South Wales.

    Vol. II   containing Maxims and Thoughts, with reflections on select passages of scripture.

    Vol. III containing Dissertations on particular subjects and occasions and Dialogues in the world of spirits.

[2] 'Active', in this context, is defined as writing letters, applying for premiums or writing articles and reports.

[3] September 1788, p. 240.

[4] This meant, apparently, that the Society would guarantee him £2,000 and then invite individual subscriptions from Members, in exchange of which it would share the Secret with them.

[5] 1788.

[6] Vol. VII.

[7] *Letters and Papers*, Vol. VII, 1795.

[8] Ibid.

[9] *Letter Book*, p. 304. November 7, 1790.

[10] In 1800 an attempt was made to obtain yet another of the King's rams, through the good offices of Sir Joseph Banks, and one arrived the following year. The 'official' experiments with the rams, carried out on behalf of the Society, were not the only ones to take place in the South-West. The 1799 volume of printed papers includes an article by Mr Parry, summarising his own experiences with the Spanish cross.

[11] This campaign against the use of children for chimney-sweeping was, in 1803, somewhat ahead of its time. Blake's poem, 'The Chimney Sweeper' appeared in 1794 in *Songs of Innocence* and Jonas Hanway had tried to draw public attention to the scandal even earlier, in 1773, but the successful campaign to end the practice was much later. Charles Kingsley's *Water Babies* was published in 1863 and Lord Shaftesbury's Act of 1875 finally put a stop to the evil.

[12] The menu at the Annual Dinner was much improved by contributions of food from wealthy members. Gifts of pineapples, grapes and venison, for instance, were made each year by Lord Lansdowne and others.

Chapter Three: EXPERIMENT AND UNCERTAINTY

[1] On this, see A. J. Peacock, *Bread or Blood: A Study of the Agrarian Riots in East Anglia in 1816*, 1965.

[2] The Mayor agreed to the Society using the site, subject to 'a small allowance in the form of a toll'.

[3] 'Claimants of the inferior class' had, it appears, become embarrassingly numerous, and at the Annual Meeting of 1808 it was resolved that only servants of Members were to be eligible for long-service awards.

⁴ In November 1807, Davis presented the Society with a portrait of his father, who had recently died.

⁵ 150 years ago, Hetling House would certainly not have presented the spick and span appearance that it does today. It appears to have been built c. 1570 by Edward Clarke, on Norman foundations and leased soon afterwards to Sir Walter Hungerford (d. 1585), of Farleigh. It subsequently became Abbey Church House and was restored in the 1950s, after air-raid damage during the 1939–45 War.

⁶ February 1808.

⁷ It is reproduced, as Appendix I, in Kenneth Hudson: *Patriotism with Profit*, Hugh Evelyn, 1972.

⁸ *Minutes*, July 1809, pp. 420–429. In March 1808, a report of the Committee of Staple Regulation endorsed all the findings relating to the Spanish cross and recommended the overall improvement of what it called 'Longram wool' by means of Government bounties.

⁹ *Letters and Papers*, Vol. XIII, 1813. In 1814 it was recorded that Thomas King, 'statuary', had presented the Society with a plaster bust of Billingsley.

¹⁰ By a decision of the Annual Meeting, 1817, the coat and buttons might also be awarded to 25 'labourers in husbandry', nominated by members.

¹¹ These buttons are now exceedingly rare. The example illustrated in the present book was given to the Society in 1972, by Mr P. Seviour, of Frome.

¹² *Minutes*, January 1819.

¹³ February 1818.

Chapter Four: INTO THE DOLDRUMS

¹ *Minutes*, 1819.

² Report to the Annual Meeting, 1819.

³ *Minutes*, December 1820.

⁴ November 1819.

⁵ *Minutes*, December 1820.

⁶ *Minutes*, December 1822.

⁷ *Minutes*, December 1819.

⁸ *Minutes*, February 1821. It is still in the Library.

⁹ *Minutes*, December 1819. Parry was awarded the Society's Gold Medal in 1820, inscribed 'The Explorer of the Polar Sea'.

¹⁰ *Minutes*, December 1819.

¹¹ *Minutes*, December 1823.

¹² *Minutes*, December 1828.

¹³ 47 Members were present.

¹⁴ An earlier premium, with the same aim, had been established in 1829, when the Bedfordean Gold Medal was offered to the farmer who 'contributed most to those dependent on him, either by letting them small portions of land for their own cultivation . . . or by any other means . . . so as to keep them from parochial aid'.

¹⁵ 1838.

¹⁶ 1837. 'The invention was thought to possess great merit.'

¹⁷ 1832. In 1834 a premium was offered for a plan of constructing suspension bridges 'in which, as in Scotland and Switzerland wire should be used instead of chains'.

¹⁸ The Literary and Scientific Institution agreed, in 1849, to offer free hospitality to the Society three times a year, and to look after its funds.

Chapter Five: THE ACLAND REVIVAL

¹ Lord Portman had become President in 1847, taking the place of Lord Lansdowne, who had agreed to become Patron, after the Royal Patronage was mysteriously withdrawn in that year.

² There was an interesting reason for the large number of Vice-Presidents. 'Those noblemen and gentlemen then acting as the Vice-Presidents of this Society were continued in office, and other noblemen and gentlemen were added to the list, with a view to preserve the balance between the western and eastern districts, without resorting to the uncourteous step of requesting any of the former Vice-Presidents to retire.'

[3] On the history of such Shows, see Kenneth Hudson: *Patriotism with Profit*, Chapter Four.

[4] Published by the Bath & West Society as a pamphlet in 1851.

[5] *Journal*, 1853, pp. 7–9.

[6] *Journal*, 1853, p. 21.

[7] *Journal*, 1853, pp. 189–190.

[8] Messrs. Acland and Pitman.

[9] *Journal*, 1853, p. 7.

[10] The Secretary spent only one year as Honorary Secretary. In 1850 he was given a salary of 50 guineas a year.

[11] An Executive Committee was appointed in 1854, mainly to look after the Annual Show.

[12] In 1852, the Bailiffs of the Town of Taunton made the Barrack Yard available for precisely three weeks.

[13] *Journal*, 1856, p. 390 has another reference to this.

[14] Sir John T. Coleridge, *Some Few Private Recollections of Sir Thomas Dyke Acland*. Privately printed, 1872. Gladstone's own copy, sent to him by the author, is in the British Museum.

[15] *Memoir and Letters of the Right Honourable Sir Thomas Dyke Acland*, edited by his son, Arthur H. D. Acland. London. Printed for private circulation, 1902.

[16] *Journal*, 1855.

[17] *Rules of the First Devon Mounted Rifle Volunteers*, together with Documents relative to the Union of Mounted and Dismounted Volunteers. Head Quarters—Broad Clyst, 1860. In Devon Record Office.

[18] *Memoir and Letters*, p. 371.

[19] *Memoir and Letters*, p. 362.

[20] Thomas Plowman: *Some Countryside Folk*. Bath: Daily Chronicle and Argus Ltd., 1911, p. 23.

[21] *Some Countryside Folk*, p. 23.

[22] The Society was greatly concerned with rural housing during the 1850s, partly, no doubt, because it was one of Acland's own special interests. See, for example, in the *Journal*, 1856, the 'Details of a Labourer's Lodging House', built for Sir Arthur Elton at Clevedon, and the plans and specifications for 'double', i.e. semi-detached, cottages, by Henry Hickes, of Bath, in the same volume.

[23] *Journal*, 1854, v.

[24] The Committee of Superintendance ceased to exist in December 1850, being replaced by the Council.

[25] *Journal*, 1854, vi.

[26] *Journal*, 1858, xv.

[27] Kenneth Hudson: *Patriotism with Profit*, p. 123.

[28] *Journal*, 1855, viii.

[29] *Journal*, 1855, p. 1.

[30] *Journal*, 1860, p. 45.

[31] In the 1861 *Journal*, however, the Society mentioned that manufacturers had become chary of exhibiting themselves, 'for fear that by doing so they might create trade jealousies, and produce difficulties in their regular routine of dealing'. They were, however, willing to send 'their best articles' to local shops, on a sale or return basis.

[32] The 'artists with West Country connections' included amateurs who were members of the Art Union.

[33] *Journal*, 1869, p. 37. The Art Committee was revived after the Second World War, and a new series of Art Exhibitions began at the Taunton Show in 1962. They are now one of the most successful features of the Show.

[34] The education project was outlined in a report to the Annual Meeting at Newton Abbot in 1857, under the heading, 'Middle Class Education'. More of Acland's ideas are to be found in his paper, 'The Education of the Farmer', which was contributed to the same volume.

[35] *Memoir and Letters*, p. 294.

[36] *Memoir and Letters*, p. 366.

[37] *Journal*, 1856, p. 388.

[38] 'The Case of the Consulting Chemists of the Royal, the Highland and the Irish Societies is quite different. When the charges here given are compared with the others, the surprise will be that such privileges can be offered to our Members at so low a rate. It is much to be desired that a fund should be raised in our Society, or that the number of members should be largely increased, so that the charges might be still further lowered.' (*Journal*, 1856, p. 392.)

[39] *Journal*, 1855, p. 62.
[40] *Journal*, 1855, p. 75.
[41] *Journal*, 1857, p. vii.
[42] *Journal*, 1862, p. 136.
[43] *Journal*, 1862, pp. 154–155.

Chapter Six: APPEALING TO A WIDER PUBLIC

[1] From E. J. T. Collins and E. L. Jones, 'Sectoral Advance in English Agriculture, 1850–80', *Agricultural History Review*, 1967.
[2] This is discussed by David Spring: *The English Landed Estate in the Nineteenth Century*, Chapter Two.
[3] F. M. L. Thompson, 'English Great Estates in the Nineteenth Century (1790–1914)', in *Contributions to the First International Conference of Economic History*, Paris, 1960, p. 394.
[4] 1859.
[5] *Journal*, 1890–91.
[6] *Journal*, 1860, v.
[7] *Journal*, 1860, v–vi.
[8] *Journal*, 1862, p. 155.
[9] *Journal*, 1857, p. 86. Tanner is described in this volume as 'Land Agent'.
[10] The Society had taken an active interest in the establishment of the Royal Agricultural College.
[11] *Journal*, 1865, pp. 243–244.
[12] *Journal*, 1869, pp. 24–25 and 35.
[13] *Journal*, 1876, pp. 48–49.
[14] *Journal*, 1878, p. 19.
[15] *Journal*, 1879, p. 28.
[16] *Journal*, 1879, p. 2.
[17] *Journal*, 1872, pp. 173–175.
[18] An article by Voelcker (*Journal*, 1880, reprinted from the *Journal of the Royal Agricultural Society*) reports that there were, at that date, twenty cheese factories in various parts of England, 'capable of dealing with the milk of about 6,000 cows'. He also mentions three factories making condensed milk, one at Aylesbury, another at Swindon, and the third at Mallow, in Ireland.
[19] *Journal*, 1863, p. 401. Report on the Exhibition of Livestock at Exeter.
[20] *Journal*, 1877, p. 29.
[21] *Journal*, 1883–84.
[22] Maule retired in 1873, but remained as Honorary Secretary. Goodwin combined the posts of Secretary and Editor of the *Journal*.
[23] *Journal*, 1877, p. 4.
[24] *Journal*, 1877.
[25] *Journal*, 1864.

Chapter Seven: THE AGE OF SHOWMANSHIP

[1] *Minutes*, 1879.
[2] *Minutes*, 1881. Dr Siemens was a notable pioneer in the use of electricity for horticulture and agriculture. The 1881 *Journal* contains a report of an address given by him to the British Association on the subject, explaining how he had been using electricity to light greenhouses, with power from a 6 h.p. steam-engine, which also heated the greenhouses. 'In order to utilise this power, I have devised means of working the dynamo-machine also during the day-time, and of transmitting the electric energy thus produced by means of wires to different points of the farm, where such operations as chaff-cutting, swede-slicing, timber-sawing and water-pumping have to be performed.'
This year the Horticultural Department is referred to for the first time as 'the Flower Tent'.
[3] Thomas Plowman: *In the Days of Victoria: Some Memories of Men and Things*. Bath: 1918.
[4] *In the Days of Victoria*, p. 11.
[5] *In the Days of Victoria*, p. 77.
[6] 6 November, 1882. This meeting was held in the GWR Board Room, at Temple Meads Station, Bristol. The notice calling it and the Agenda were sent out, in the normal way, from the Society's

offices at 4 Terrace Walk, Bath. Special meetings of the Society were not infrequently held in the Board Room at Temple Meads and also at the Great Western Hotel, Paddington.

[7] *Journal*, 1904–5, p. 105.

[8] *Journal*, 1904–5, p. 106.

[9] 18 August, 1906.

[10] At Brighton, in 1886, a different system was adopted, the Society managing the Working Dairy itself. On this occasion, the separator was driven by a horse-gear.

[11] *Journal*, 1890–1.

[12] *Journal*, 1890–1.

[13] *Journal*, 1884–5.

[14] Voelcker had recently died. The 1884–5 volume contains an obituary and appreciation of him, by Sir T. D. Acland. His son was appointed to succeed him as the Society's Consultant Chemist. A consultant Botanist, W. Carruthers, FRS, was appointed in 1887.

[15] *Journal*, 1887.

[16] *Journal*, 1886.

[17] *Journal*, 1888, p. 4.

[18] This had in fact been customary at the Society's Christmas shows in the early years.

[19] *Journal*, 1890, p. 83. This was, of course, the son of the original Augustus Voelcker. He appears to have been a somewhat touchy man. In 1894 he objected to not being consulted over the Society's employment of another chemist for work connected with the Fertilizers and Feeding Stuffs Act, and to the publication of the results of his rival's work in the *Journal*.

[20] *Journal*, 1892–3.

[21] *Journal*, 1892–3.

[22] *Journal*, 1883–4.

[23] *Journal*, 1891–2.

[24] *Journal*, 1892–3.

[25] Cornwall, Devon, Somerset, Dorset, Wiltshire, Gloucestershire, Herefordshire, Worcestershire, Monmouthshire, South Wales.

[26] Oxfordshire, Berkshire, Hampshire, Surrey, Sussex, Kent.

[27] At Leeds, for Yorkshire; at Newcastle-on-Tyne, for Durham and Northumberland; at Aberystwyth, for mid-Wales.

[28] *Journal*, 1896–7.

[29] *Journal*, 1898–9. A Standing Committee was set up, to carry the proposals into effect.

Chapter Eight: FOREIGN COMPETITION AND WAR AS INCENTIVES TO EFFICIENCY

[1] 'The Sheep Stocks of the World'.

[2] Islington was chosen as an example.

[3] Of the famous bacon firm at Calne, in Wiltshire.

[4] *Journal*, 1910–11.

[5] James Long, *Journal*, 1914–15.

[6] See, for instance, Bastin's paper (1912–13) on wheat pests, with first-class photographs. Another remarkable paper by him, 'On stimulating the growth of plants by means of applied electricity', was published in the 1909–10 volume.

[7] *Journal*, 1910–11.

[8] *Journal*, 1911–12.

[9] 'How the dairy industry has progressed,' *Journal*, 1914–15.

[10] Departures were as sad and serious as deaths. Particularly regrettable was that of the Consultant Botanist, J. H. Priestley, who left in 1912, 'on his promotion to an important post in the North of England'.

[11] *Journal*, 1916–17. In 1908 Gibbons laid before the Society detailed proposals for holding an Autumn Show. These, however, were not accepted.

[12] The present author had the benefit of several conversations with Mr Ayre during 1968, at his home in Limpley Stoke. He died three years later, at the age of 93.

[13] In a conversation with the author, August 1968.

[14] He disapproved of the Society's permanent site at Shepton Mallet for this and other reasons.

[15] These occurred long before Ayre's time. At the Stewards' meeting in 1903 a letter was considered from Clifton Golf Club, asking the Society, 'as the Club had been put to inconvenience, owing to the Show occupying part of their links, to present the Club with a Champion Cup'.

[16] *Journal*, 1889.
[17] *Journal*, 1905–6.
[18] *Journal*, 1905–6.
[19] *Journal*, 1911–12.
[20] Since 1904 it had been installed in new offices at 3 Pierrepont Street, Bath. The house had been bought and furnished 'without any call having been made upon the reserve fund' and the accommodation was much more convenient than its predecessors, from Hetling House onwards.
[21] *Journal*, 1915–16.
[22] *Journal*, 1915–16.
[23] *Journal*, 1901–2. At the Croydon Show in 1901 there was an exhibition, arranged by Lumley and Co., to illustrate the processes of drying and evaporating fruit and vegetables.
[24] *Journal*, 1915–16.
[25] *Journal*, 1904–5. 'The Education of the Labourer.'
[26] It was appropriate that Plowman should write it, since Acland, like his father, had been the Chairman of the *Journal* Committee. He was also one of the Society's Trustees, and had been a Member since 1872.
[27] *Journal*, 1918–19.

Chapter Nine: DEPRESSION AND DULLNESS BETWEEN THE WARS

[1] Finance Committee Minutes, 1919.
[2] This was altered later to £650 net. The Assistant Secretary received £450 from June 1920, together with a bonus of £100 for work undertaken during the late Secretary's illness. Plowman was voted an allowance of £100 a year, as from 1 October, 1919. (He had been ill since June). After his death later that year, Mrs Plowman was given a grant of £50, and an annual allowance of £100 in recognition of her late husband's services as Secretary and Editor.
[3] *Journal*, 1920–21.
[4] *Journal*, 1921–22.
[5] As the result of a Council meeting held on 27 July, 1920.
[6] *Journal*, 1922–23. The new policy does not seem to have produced any very obvious economy. As late as 1929 the Long Ashton report still required seventy-seven pages of the *Journal*.
[7] *Journal*, 1923–24.
[8] *Journal*, 1926–27.
[9] This was also the first Meeting to be held after the Society had absorbed the Somerset County Agricultural Association. The amalgamation brought the total membership up to 1,300.
[10] *Journal*, 1927–28.
[11] *Journal*, 1928–29.
[12] *Journal*, 1929–30.
[13] *Journal*, 1930–31.
[14] *Journal*, 1937–38.
[15] *Journal*, 1933–34. The special committee appointed in 1928 to look carefully into expenditure on the Show gave rise to the optimistic belief that the Society could look forward to 'increased efficiency and the ability to remain unaffected by the depression which has affected other Show Societies'. (*Journal*, 1928–29.)
[16] In recognition of his services the University conferred the honorary degree of Doctor of Laws on him in 1932.
[17] *Journal*, 1935–36.
[18] Conversation with the author, 25 February, 1969. Yardley came to Bath in 1937. The post had been advertised in 1935, but the appointment was delayed because the Society was trying to amalgamate with the Royal Counties.
[19] Ibid.
[20] The pamphlets which appeared during the war years were:
A. W. Ling: *The growing of flax, linseed, sugar beet and potatoes*, 1940.
Sir R. G. Stapledon: *The utilisation of our grassland areas*, 1941.
A. W. Ling and J. W. Egdell: *Milk production in war-time*, 1941.
J. A. Scott Watson: *Farming after the war*, 1941.
E. E. Edwards and T. I. Davies: *Pests of ploughed-up grassland*, 1942.
W. T. Price: *The agricultural unit in relation to the small farm*, 1942.
A. Bridges: *Farm accounts for income tax purposes*, 1942.

F. A. Bush: *Gardens and allotments in war-time*, 1943.
C. P. Ackers: *Forestry*, 1943.
B. T. P. Barker: *Agricultural and horticultural research in war-time*, 1943.
S. J. Wright: *Use and development of agricultural machinery during the war*, 1944.
J. Hammond: *The improvement of cattle*, 1944.
E. M. Crowther: *Fertilisers during the war and after*, 1945.

Chapter Ten: A PERMANENT HOME

[1] The previous highest total had been 115,000 at Swansea in 1923.
[2] Conversation with the author, 25 February, 1969.
[3] *Annual Report*, 1948.
[4] By making a direct approach to General Horrocks, Yardley got the Household Cavalry for nothing in 1948, although he had to pay in subsequent years.
[5] *Annual Report*, 1948.
[6] This had been deliberately encouraged by Yardley, in order to attract as many visitors as possible. There had been a large increase in membership in 1947, the year of the Cheltenham Show, when ladies were offered membership at 10s, which was less than the price of a season ticket.
[7] One additional clerk and one additional typist eventually proved sufficient.
[8] *Annual Report*, 1948.
[9] *Annual Report*, 1949.
[10] *Annual Report*, 1949.
[11] Western Division (Devon and Cornwall); Central Division (Somerset, Dorset and Wiltshire); Southern Division (Hampshire, Berkshire, Oxfordshire, Buckinghamshire, Middlesex, Surrey, Sussex, Kent); North-Western Division (Worcestershire, Gloucestershire, Herefordshire, Monmouthshire and Wales).
[12] *Annual Report*, 1951.
[13] Some indication of the state of the Society's finances at this time is given in a Minute of the Finance and General Purposes Committee in September 1952. The Committee felt obliged to refuse a request to contribute to the funds of the Museum of English Rural Life at Reading and to confine its subscriptions to the following: Animal Health Trust (£10 10s 0d); Association of Agriculture (£10 10s 0d); British Field Sports Society (£3 3s 0d); British Horse Society (£1 1s 0d); British Rabbit Council (10s); British Show Jumping Association (£9); Hunters' Improvement Society (£1 1s 0d); Livestock Export Group (£2 2s 0d); National Pony Society (£1 1s 0d); Poultry Club (£1 1s 0d); Severn Wildfowl Trust (£1 1s 0d); Shire Horse Society (£2 2s 0d); Show and Breed Secretaries' Association (£1 1s 0d); Hackney Horse Society (£1 1s 0d).
[14] *Annual Report*, 1952. The 1952 Show had an unusual number of visitors from abroad. 'The Society,' it was reported, 'had the pleasure of entertaining to lunch 50 Australian Farmers and a party of 30 Agriculturalists from Denmark.'
[15] *Annual Report*, 1953.
[16] *Annual Report*, 1952.
[17] *Annual Report*, 1955.
[18] *Annual Report*, 1957.
[19] *Annual Report*, 1957.
[20] *Annual Report*, 1959.
[21] *Annual Report*, 1960.
[22] *Annual Report*, 1961.
[23] The association between the Royal Smithfield Club and the Bath & West Society was terminated on 1 July, 1961.
[24] *Annual Report*, 1963.
[25] *Annual Report*, 1964.
[26] These were used for the 1969 Somerset Cattle Breeding Centre's Show and Sale.
[27] At the Annual General Meeting, 1969.
[28] *Annual Report*, 1971.
[29] £30,000 in 1964.
[30] Partly financed by the National Westminster Bank.
[31] John Davis's services to the Society and to agriculture were recognised by the award to him of an MBE in the 1973 New Year's Honours List.
[32] *Annual Report*, 1973.

[33] One uses the title only to preserve a sense of continuity. Yesterday's Secretary is now, since 1974, the Society's Chief Executive and yesterday's Assistant Secretary is now the Secretary, concerned mainly with the Show.

Appendix Two: SECRETARIES AND PRESIDENTS OF THE SOCIETY

[1] From 1865 until 1870 Maule was referred to as 'Hon. Sec.'.
[2] The new system of holding the Annual Meeting at the Summer Show, instead of in December, began in 1852 and from then on there were annual Presidents, whose tenure ran from October.

Appendix Three: MEMBERSHIP, INCOME AND EXPENDITURE

[1] Comparable figures for earlier decades are not available.

Appendix Four: PLACES WHERE THE ANNUAL SHOW WAS HELD

[1] No figures are available until 1860, which may well have been the year when that useful invention, the automatically-registering turnstile, first became available.

Index